Advanced Manufacturing Processes

The field of manufacturing science has evolved over the years with the introduction of non-traditional machining processes. This reference book introduces the latest trends in modeling and optimization of manufacturing processes.

It comprehensively covers important topics including additive manufacturing at multi-scales, sustainable manufacturing, rapid manufacturing of metallic components using 3D printing, ultrasonic-assisted bone drilling for biomedical applications, micromachining, and laser-assisted machining.

This book is useful to senior undergraduate and graduate students in the fields of mechanical engineering, industrial and production engineering, and aerospace engineering.

Mathematical Engineering, Manufacturing, and Management Sciences

Series Editor
Mangey Ram
Professor, Assistant Dean (International Affairs),
Department of Mathematics, Graphic Era University, Dehradun, India

The aim of this new book series is to publish the research studies and articles that bring up the latest development and research applied to mathematics and its applications in the manufacturing and management sciences areas. Mathematical tool and techniques are the strength of engineering sciences. They form the common foundation of all novel disciplines as engineering evolves and develops. The series will include a comprehensive range of applied mathematics and its application in engineering areas such as optimization techniques, mathematical modelling and simulation, stochastic processes and systems engineering, safety-critical system performance, system safety, system security, high assurance software architecture and design, mathematical modelling in environmental safety sciences, finite element methods, differential equations, reliability engineering, etc.

Differential Equations in Engineering
Research and Applications
Edited by Nupur Goyal, Piotr Kulczycki, and Mangey Ram

Sustainability in Industry 4.0
Challenges and Remedies
Edited by Shwetank Avikal, Amit Raj Singh, Mangey Ram

Applied Mathematical Modeling and Analysis in Renewable Energy
Edited by Manoj Sahni and Ritu Sahni

Swarm Intelligence: Foundation, Principles, and Engineering Applications
Abhishek Sharma, Abhinav Sharma, Jitendra Kumar Pandey, and Mangey Ram

Advances in Sustainable Machining and Manufacturing Processes
Edited by Kishor Kumar Gajrani, Arbind Prasad and Ashwani Kumar

Advanced Materials for Biomechanical Applications
Edited by Ashwani Kumar, Mangey Ram and Yogesh Kumar Singla

Biodegradable Composites for Packaging Applications
Edited by Arbind Prasad, Ashwini Kumar and Kishor Kumar Gajrani

Computing and Stimulation for Engineers
Edited by Ziya Uddin, Mukesh Kumar Awasthi, Rishi Asthana and Mangey Ram

Advanced Manufacturing Processes
Edited by Yashvir Singh, Nishant K. Singh and Mangey Ram

For more information about this series, please visit: www.routledge.com/Mathematical-Engineering-Manufacturing-and-Management-Sciences/book-series/CRCMEMMS

Advanced Manufacturing Processes

Edited by
Yashvir Singh
Nishant K. Singh
Mangey Ram

CRC Press is an imprint of the
Taylor & Francis Group, an **informa** business

MATLAB® is a trademark of The MathWorks, Inc. and is used with permission. The MathWorks does not warrant the accuracy of the text or exercises in this book. This book's use or discussion of MATLAB® software or related products does not constitute endorsement or sponsorship by The MathWorks of a particular pedagogical approach or particular use of the MATLAB® software.

First edition published 2023
by CRC Press
6000 Broken Sound Parkway NW, Suite 300, Boca Raton, FL 33487–2742

and by CRC Press
4 Park Square, Milton Park, Abingdon, Oxon, OX14 4RN

CRC Press is an imprint of Taylor & Francis Group, LLC

© 2023 selection and editorial matter, Yashvir Singh, Nishant K. Singh, Mangey Ram; individual chapters, the contributors

Reasonable efforts have been made to publish reliable data and information, but the author and publisher cannot assume responsibility for the validity of all materials or the consequences of their use. The authors and publishers have attempted to trace the copyright holders of all material reproduced in this publication and apologize to copyright holders if permission to publish in this form has not been obtained. If any copyright material has not been acknowledged please write and let us know so we may rectify in any future reprint.

Except as permitted under U.S. Copyright Law, no part of this book may be reprinted, reproduced, transmitted, or utilized in any form by any electronic, mechanical, or other means, now known or hereafter invented, including photocopying, microfilming, and recording, or in any information storage or retrieval system, without written permission from the publishers.

For permission to photocopy or use material electronically from this work, access *www.copyright.com* or contact the Copyright Clearance Center, Inc. (CCC), 222 Rosewood Drive, Danvers, MA 01923, 978–750–8400. For works that are not available on CCC please contact *mpkbookspermissions@tandf.co.uk*

Trademark notice: Product or corporate names may be trademarks or registered trademarks and are used only for identification and explanation without intent to infringe.

ISBN: 978-1-032-05446-9 (hbk)
ISBN: 978-1-032-11513-9 (pbk)
ISBN: 978-1-003-22023-7 (ebk)

DOI: 10.1201/9781003220237

Typeset in Times
by Apex CoVantage, LLC

Contents

Preface vii

Editors ix

Contributors xi

Chapter 1   Recent Trends of Cutting Fluids and Lubrication
Techniques in Machining 1

Ramandeep Singh and Varun Sharma

Chapter 2   Electrical Discharge Drilling of Al-SiC Composite
with Gas-Assisted Multi-Hole Rotary Slotted Tool 29

*Nishant K. Singh, Yashvir Singh, Bholey Singh,
and Hardik Berwal*

Chapter 3   Web Buckling Investigation of Direct Metal Laser Sintering-
Based Connecting Rod with Hexagonal Perforations 51

*Gulam Mohammed Sayeed Ahmed, Mengistu Gelaw Perumall,
Janaki Ramulu, Belay Brehane, Devendra Kumar Sinha, and
Satyam Shivam Gautam*

Chapter 4   Materials for Additive Manufacturing: Concept,
Technologies, Applications and Advancements 79

M. Anugrahaprada and Pawan Sharma

Chapter 5   Green 3D Printing: Advancement to Sustainable
Manufacturing 99

*Amber Batwara, Harsh Mundra, Apoorva Daga,
Vikram Sharma, and Mohit Makkar*

Chapter 6   Nanotechnology and Manufacturing 119

*Vijay K. Singh, Puneet Kumar, Manikant Paswan,
T. Ch. Anil Kumar, and Saipad B.B.P.J. Sahu*

Chapter 7   Modeling and Optimization of Process Parameters with
Single Point Incremental Forming of AA6061 Using
Response Surface Method 129

*Assefa Leramo, Devendra Kumar Sinha, and
Satyam Shivam Gautam*

v

Chapter 8 Plasma Fundamentals for Processing of Advanced Materials 147

Tapan Dash and Bijan Bihari Nayak

Chapter 9 The Concept of Rotary Ultrasonic Bone Machining during Orthopaedic Surgeries 157

Raj Agarwal, Jaskaran Singh, Vishal Gupta, and Ravinder Pal Singh

Chapter 10 Advancement in Magnetic Field Assisted Finishing Processes 173

Girish C. Verma, Dayanidhi K. Pathak, Pawan Sharma, Aviral Misra, and Pulak M. Pandey

Index 195

Preface

Recent trends in advanced machining are perhaps the most challenging and demanding field of research. In this first book to address all major technologies in this area, users can find the latest technological innovations and analysis in one spot, making it easy to compare performance requirements. The technology solutions encompassed tend to involve mechanical, thermal, chemical, micro and hybrid machining processes, and the highly innovative finishing techniques. Advanced manufacturing processes bring on scientific advances in materials science, conventional manufacturing processes, rapidly growing additive/hybrid technologies, and the industrial Internet of Things to develop realistic solutions for industry. The goal of this book is to compile original research articles outlining the recent developments in advanced manufacturing technology such as additive manufacturing, materials and manufacturing process production operations, and manufacturing sustainability. This book will have tremendous relevance in the lives of academic institutions, practitioners, researchers, and industry players.

This book is unique because it covers a wide array of modern machining methodologies; provides a description of potential techniques, including the components and materials required; keeps readers apprised of recent developments in nanotechnology application in manufacturing; is based on the most recent developments in advanced sustainable manufacturing processes; points to new areas for further work such as rapid manufacturing of components using 3D printing; shares detailed knowledge of applications of magnetic field assisted polishing processes for super finishing of complex surfaces; discusses ultrasonic-assisted drilling for biomedical application; and covers recent trends in modeling and optimization of manufacturing processes. This book presents pertinent and realistic global interpretations of advanced manufacturing processes and provides the latest research progress in the field of advanced manufacturing processes in engineering sciences. In addition, the book certainly serves as a guide for the implementations of the latest engineering technologies, providing theoretically sound context with relevant case studies.

MATLAB® is a registered trademark of The Math Works, Inc.
For product information, please contact:
The Math Works, Inc.
3 Apple Hill Drive
Natick, MA 01760-2098
Tel: 508-647-7000
Fax: 508-647-7001
E-mail: info@mathworks.com
Web: http://www.mathworks.com

Editors

Dr. Yashvir Singh is presently working as an associate professor in the Department of Mechanical Engineering, Graphic Era Deemed to be University, Dehradun, Uttarakhand, India. He has more than 15 years of teaching experience. He has written more than 85 research articles and published in various peer-reviewed journals. He is also a reviewer and editorial board member of various journals. His specialization includes areas like tribology, biofuels, lubrication, manufacturing, etc.

Dr. Nishant K. Singh is an associate professor at Hindustan College of Science and Technology, Mathura, Uttar Pradesh, India. He earned his PhD from IIT, Dhanbad, with a master's in production engineering and a BTech in mechanical engineering from the Delhi College of Engineering. He has written more than 40 articles in well-known international journals and serves as a reviewer and editor of peer-reviewed journals and conferences. His research interests include tribology, micro-manufacturing, and non-conventional machining processes.

Dr. Mangey Ram received his PhD with a major in mathematics and minor in computer science from G. B. Pant University of Agriculture and Technology, Pantnagar, in 2008. He is an editorial board member of many international journals. He has published 102 research publications in national and international journals of repute. His fields of research are operations research, reliability theory, fuzzy reliability, and systems engineering. Currently, he is working as a professor at Graphic Era University.

Contributors

Raj Agarwal
Mechanical Engineering Department
Thapar Institute of Engineering and
   Technology
Patiala, Punjab, India

Gulam Mohammed Sayeed Ahmed
Center of Excellence for Advanced
   Manufacturing Engineering
Program of Mechanical Design and
   Manufacturing Engineering
School of Mechanical, Chemical and
   Materials Engineering (SoMCME)
Adama Science and Technology
   University
Adama, Ethiopia

M. Anugrahaprada
Department of Mechanical
   Engineering
Sardar Vallabhbhai National Institute
   of Technology Surat
Gujarat, India

Amber Batwara
Department of Mechanical
   Mechatronics Engineering
LNM Institute of Information
   Technology
Jaipur, India

Hardik Berwal
Department of Production Engineering
Birla Vishvakarma Mahavidhyalaya
Anand, Gujrat, India

Belay Brehane
School of Mechanical, Chemical &
   Materials Eng.
Adama Science and Technology
   University
Adama, Ethiopia

Apoorva Daga
Department of Mechanical
   Mechatronics Engineering
LNM Institute of Information
   Technology
Jaipur, India

Tapan Dash
Centurion University of Technology and
   Management
Odisha, India

Satyam Shivam Gautam
Department of Mechanical Engineering
North Eastern Regional Institute of
   Science and Technology
Itanagar, Arunachal Pradesh, India

Vishal Gupta
Mechanical Engineering Department
Thapar Institute of Engineering and
   Technology
Patiala, Punjab, India

Puneet Kumar
Karunya Institute of Technology and
   Sciences, Deemed to be University
Karunya Nagar, Coimbatore, Tamil
   Nadu, India

T. Ch. Anil Kumar
Vignan's Foundation for Science
   Technology and Research
Vadlamudi, Guntur Dt., India
Andhra Pradesh, India

Assefa Leramo
Department of Mechanical Design and
   Manufacturing Engineering
Adama Science and Technology
   University
Adama, Ethiopia

Mohit Makkar
Department of Mechanical
  Mechatronics Engineering
LNM Institute of Information
  Technology
Jaipur, India

Aviral Misra
Industrial and Production Engineering
  Department
Dr. B. R. Ambedkar National Institute
  of Technology
Jalandhar, Punjab, India

Harsh Mundra
Department of Mechanical
  Mechatronics Engineering,
LNM Institute of Information
  Technology
Jaipur, India

Bijan Bihari Nayak
CSIR-Institute of Minerals and
  Materials Technology
Bhubaneswar, India

Pulak M. Pandey
Department of Mechanical
  Engineering
Indian Institute of
  Technology
Delhi, New Delhi, India

Manikant Paswan
National Institute of
  Technology
Jamshedpur, Jharkhand, India

Dayanidhi K. Pathak
Mechanical and Automation
  Engineering Department
G. B. Pant Government Engineering
  College Delhi
New Delhi, India

Mengistu Gelaw Perumall
Department of Mechanical Design and
  Manufacturing Engineering
School of Mechanical, Chemical and
  Materials Engineering
Adama Science and Technology
  University
Adama, Ethiopia

Janaki Ramulu
Center of Excellence for Advanced
  Manufacturing Engineering
Program of Mechanical Design and
  Manufacturing Engineering
School of Mechanical, Chemical
  and Materials Engineering
  (SoMCME)
Adama Science and Technology
  University
Adama, Ethiopia

Saipad B.B.P.J. Sahu
Gita Autonomous College
Bhubaneswar, Dt. Khurda, Odisha,
  India

Pawan Sharma
Department of Mechanical
  Engineering
Indian Institute of Technology (BHU)
Varanasi, Uttar Pradesh, India

Varun Sharma
Department of Mechanical and
  Industrial Engineering
Indian Institute of Technology
Roorkee, India

Vikram Sharma
Department of Mechanical
  Mechatronics Engineering
LNM Institute of Information
  Technology
Jaipur, India

Contributors

Bholey Singh
Department of Mathematics
G. L. Bajaj Group of Institutions
Mathura, UP, India

Jaskaran Singh
Mechanical Engineering
  Department
Thapar Institute of Engineering and
  Technology
Patiala, Punjab, India

Nishant K. Singh
Hindustan College of Science and
  Technology
Mathura, UP, India

Ramandeep Singh
Department of Mechanical and
  Industrial Engineering
Indian Institute of Technology
Roorkee, India

Ravinder Pal Singh
Department of Mechanical Engineering,
  Maharishi Markandeshwar (Deemed
  to be University)
Mullana, India

Vijay K. Singh
Madan Mohan Malaviya University of
  Technology
Gorakhpur, India

Yashvir Singh
Department of Mechanical Engineering
Graphic Era Deemed to be University
Dehradun, Uttarakhand, India

Devendra Kumar Sinha
Center of Excellence for
  Advanced Manufacturing
  Engineering
Program of Mechanical Design
  and Manufacturing
  Engineering
School of Mechanical, Chemical
  and Materials Engineering
  (SoMCME)
Adama Science and Technology
  University
Adama, Ethiopia

Girish C. Verma
Mechanical Engineering Department
Indian Institute of Technology
Indore, India

1 Recent Trends of Cutting Fluids and Lubrication Techniques in Machining

Ramandeep Singh and Varun Sharma

1.1	Introduction	1
1.2	Classification of Cutting Fluids	4
	1.2.1 Conventional Cutting Fluids	4
	1.2.2 Vegetable Oil-Based Cutting Fluids	5
	1.2.3 Nanofluids	6
1.3	Techniques of Cleaner Machining	8
	1.3.1 Conventional Cooling	9
	1.3.2 Non-Conventional Cooling	9
	1.3.2.1 Dry Machining	9
	1.3.2.2 Minimum Quantity Lubrication (MQL)	10
	1.3.2.3 Cryogenic Cooling	12
	1.3.2.4 Ultrasonic Atomization of Cutting Fluid	15
1.4	Case Study	16
	1.4.1 Stability Study	17
	1.4.2 Characterization of Nanofluids	19
	1.4.3 Tribological Study	20
	1.4.4 Machining Study	21
1.5	Conclusions	24
	References	25

1.1 INTRODUCTION

Machining is an essential process in the manufacturing cycle of a product. It is a process of removing unwanted material from the workpiece through chip formation. This is generally performed at the later stage of the manufacturing cycle, which can put forward its significance for the life of the machined components. The machining process is used extensively for the production of necessary component size and shape. The component could be machined at high speed for improving productivity. However, a lot of heat is produced during machining because of the friction at the tool and workpiece contact pairs. This, in turn, results in high temperature, high pressure, friction, and considerable tool wear. Also, it results in the creation of tensile residual stresses and chip segmentation and affects the workpiece's surface integrity. Hence, there is a need of cutting fluid to reduce the temperature of the cutting zone

DOI: 10.1201/9781003220237-1

and provide sufficient lubrication to reduce the friction and for evacuation of chips from the machining zone [1–4].

The use of lubricants in the metal cutting industry started from the industrial revolution in the late 18th century. Generally, the metal cutting fluids can be straight oils or some soluble oils. Straight oils are those which are used directly without adding any additive in them while soluble oils are ones that are mixed with some other additive and are soluble in water or other liquids [5–6].

Cutting fluid performs microcapillary action in the machining zone to provide sufficient lubrication, acts as a heat-carrying medium, and also helps in chip evacuation [7]. Cutting fluids are generally used for difficult-to-machine materials and in complex machining. Metalworking processes are intended to have tool-work contact [8]. In order to separate them, a boundary of lubricant needs to be placed in between. The problem of friction and tool wear is addressed by the lubricating capability of the cutting fluid while the cooling action encounters high temperatures. The various factors such as cutting conditions, cutting environment, cutting tool, machine tool chatter, and workpiece nature significantly influence machining performance. The cutting fluid, its thermo-fluidic properties, method of application, its flow rate/pressure, along with quantity, is, in turn, directly linked with the machining performance. The factors that control the machining performance are shown in Figure 1.1.

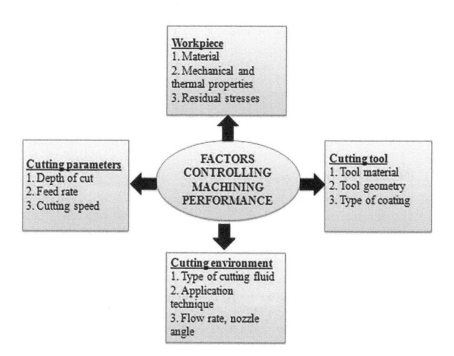

FIGURE 1.1 Depiction of factors that control machining performance.

Cutting Fluids and Lubrication Techniques

Metalworking fluids are used in order to enhance the surface integrity of parts, increased tool life, etc. However, cutting fluids cost nearly 17 percent of the manufacturing cost. Besides this, almost all the conventional cutting fluids are prepared by mixing with water, increasing water consumption. Apart from the economic perspective, cutting fluids negatively impacts the environment through contamination of soil, polluting air and water. Besides this, cutting fluids creates serious health issues for the metalworking industry, like ingestion, toxicity, inhalation, and skin irritation. The National Institute of Occupational Safety and Health (NIOSH) [9–10] report shows oil-based cutting fluids causing ingestion, irritation, and inhalation, which creates several health issues among the workers. The various ways through which cutting fluid could invade humans are mentioned in Figure 1.2.

Therefore, there is a need to reduce the cutting fluid consumption from an economic point of view and create environment and human-friendly cutting fluid to address the sustainability aspect of the metal cutting industry.

Conventional lubricants used in the machining processes were mainly derived from mineral oils and petroleum-based oils. However, in recent years, many research attempts have been made to prepare cutting fluids from vegetable oils, which has resulted in better machining performance. The chemically modified vegetable oils

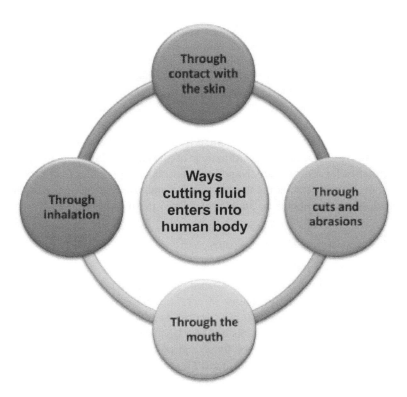

FIGURE 1.2 The common paths through which metalworking fluids intrude into humans [11].

can work as a lubricant in the machining process due to good thermo-physical properties. However, their use is limited due to poor thermal stability and oxidation at high temperatures. However, researchers are looking for suitable alternatives with the growing awareness regarding the environmental threat and health problems associated with cutting fluids.

The mixing of additives has also been found to enhance the properties of vegetable oils and make them suitable for high-temperature applications. The additives like ionic liquids, nanoparticles, nanotubes, nanofibres, and surfactants have proved to enhance the base oil's thermo-physical properties, which in turn improves the lubrication performance.

Therefore, to address the sustainability aspect in the machining industry, research needs to be performed to develop sustainable cutting fluids and produce sustainable techniques of cutting fluid applications.

Previous literature is focused more on the techniques of lubrication and cutting fluids. This work is dedicated to developing various cutting fluids, the evolution of lubrication techniques, application methods, and their challenges. A case study of developing a novel cutting fluid and its application in machining has been discussed at the end of the chapter.

1.2 CLASSIFICATION OF CUTTING FLUIDS

Several types of cutting fluids have been employed in the machining industry so far. Lubricants used as cutting fluids are mainly solid, liquid, and gaseous, but liquid lubricants are primarily used. The liquid lubricants can be classified based on their origin. Conventional fluids are obtained from petroleum-based mineral oils, which are extracted from crude oils. Conventional cutting fluids used are mainly petroleum-based synthetic lubricants like paraffin, kerosene, SAE 80, etc. Due to the problems associated with these lubricants, the use of biodegradable oil has become popular. Vegetable oil-based lubricants are obtained from plants and create less threat to the environment. To enhance the performance of the cutting fluids, nanofluids have been in practice for recent years. Nanofluids have become popular in the machining industry due to their outstanding thermo-physical and tribological behavior [11]. The detailed classification of cutting fluids is explained in this section.

1.2.1 CONVENTIONAL CUTTING FLUIDS

Conventional cutting fluids are mineral oils, and synthetic oils made up of non-petroleum manufactured base stock. Mineral oils are generally obtained from crude oil and consist of hydrocarbon chains. Mineral oils have the molecular structure of cyclic carbon chains. The mineral oils can be straight, fatty oils, and soluble oils. The straight oils are low viscous oils used for light-duty machining operations. The fatty oils are used for better surface finish and precision as they provide good wettability between tool-workpiece interfaces. The soluble oils are a mixture of mineral oils with emulsifiers. They provide lubrication due to the oil presence and cooling action because of water presence. Mineral oils are cheap and are used for several applications [6].

However, there are several limitations of mineral oils, and there is a need to develop other kinds of cutting fluids. The major problem with mineral oils is their inadequate cooling capacity, which reduces their chances of working in high-speed machining. Also, it causes more wearing of the tool, thereby affecting the surface integrity of the material. Furthermore, the mineral oils are non-biodegradable, and their toxicity creates a threat to the environment and occupational health. The growing concern of the environment has reduced the use of mineral oils as cutting fluids. This has paved research for the development of green cutting fluids [12].

Synthetic cutting fluids were manufactured to overcome the mineral oil limitations. Synthetic oils constitute discrete compounds and are different as compared to oils derived from petroleum. They involve compounds like Poly Alkylene Glycol (PAG), synthetic and ester-based hydrocarbons. Synthetic oils are less toxic, have low vapour pressure, and good thermal stability makes them suitable for working in high-temperature machining operations. Synthetic lubricants can be manufactured for use in specific applications involving high temperature and high pressure. Synthetic oils are water-miscible and provide good wetting, non-corrosiveness, and good lubricity.

However, synthetic oils also have several issues related to their disposal and health-related issues because of cutting fluid's synthetic nature. For example, synthetic cutting fluid can contact the skin and create skin irritation and other diseases. Also, their high limits the use of synthetic oils. As a result, researchers have focused on manufacturing cutting fluids from plant-based oils to address these issues of traditional cutting fluids.

1.2.2 Vegetable Oil-Based Cutting Fluids

The growing concern of the environment and government regulations have forced industries to use renewable sources. This has encouraged the metalworking industries to use vegetables as the base oil of a lubricant. Vegetable and plant-based oils consisted of long-chain fatty acids with unsaturated double bonds that provide lubrication properties [13]. These oils have high fire and flashpoints, high viscosity index like traditional cutting fluids. At the same time, they are biocompatible, renewable, less toxic, and economical [14–15].

The environment-friendly cutting fluids usually contain pristine vegetable oils and fluids that are chemically synthesized like esters involving phosphate-based esters, alkylated aromatics, polybutanes, polyalphaolefins, and polyalklene glycols. The major vegetable-based oils like soybean oil, Jatropha curcas oil, canola oil, rapeseed oil, coconut oil, sunflower oil, hazelnut oil, neem oil, castor oil, and olive oil have been used as a lubricant in machining. They are found to improve the cutting performance, life of the tool, and surface finishing compared to petrochemical cutting fluids [16–18].

In this regard, three types of vegetable oils (hydrotreated rapeseed oil, rapeseed methyl ester oil, and Jatropha curcas L. oil) have been investigated for their thermophysical and lubrication characteristics. It has been found that Jatropha curcas oil is found to have better thermophysical properties (dynamic viscosity, density, fire point, and total acid number) along with superior tribological properties (wear and friction) than the other two oils [19]. Furthermore, the effects of jatropha and moringa oil as cutting fluid for the machining of aluminium alloys were investigated by Souza et al. [20]. It was observed that the thermo-physical and tribological behaviour of jatropha and moringa oil had been far off than the conventional mineral oil. Also, they found that the jatropha oil showed the best results during the machining process.

In this regard, the blending of different vegetable oils was investigated for their suitability as cutting fluids in mild steel drilling operations. In the work reported by Susmitha et al. [21], different vegetable oils (non-edible) such as Neem oil and Karanja oil (Honge) along with a blend of these two oils were used. The influence of these cutting fluids over cutting force, formation of chips, and roughness of surface are investigated. The obtained results were compared with petroleum-based cutting fluids and dry machining. The cutting fluid prepared after blending 50-percent Neem–50-percent Karanja gives the best specific heat, high flash, optimum dynamic viscosity, adhesiveness, and fire point. The continuous chips were uniformly formed in the case of 50-percent Karanja–50-percent Neem blend. These continuous chips produced in the machining of mild steel for the blend of 50-percent Karanja–50-percent Neem had uncoloured silver showing less heat carried by the chip. Other than that, a reduction in machining force by 169.23N for the 50-percent Karanja–50-percent Neem blend has also been reported compared to dry cutting and least among other fluids.

However, vegetable oils as a neat lubricant are not preferred due to low oxidative stability, low corrosion resistance, and poor tribological behaviour [22]. The characteristics of vegetable oils can be enhanced using additives and surfactants [13]. Mixing additives like ionic liquids, nanoparticles, and surfactants into vegetable oils has displayed better lubrication properties.

1.2.3 NANOFLUIDS

Nanofluids belong to that class of fluids that are prepared by dispersing nano-sized materials (nanoparticles, nanofibers, and nanotubes) into the base fluid. When added into the base oils, nanoparticles create Brownian motion, which enhances the stability and thermo-physical behavior of the base fluid. As a result, nanofluids possess excellent thermo-physical attributes such as convective heat transfer coefficient, viscosity, and thermal conductivity compared to base fluids. Thus, nanofluids have broader relevance in enhancing heat transfer during several applications in industries, electronics, food, biomedicine, transportation, and nuclear reactors [23–26].

These fluids are classified as unitary nanofluid and hybrid nanofluid. Unitary nanofluid includes only one type of nanoparticles dispersed in the base oil, whereas hybrid nanofluids consist of mixing of two or more nanoparticles into the base oil. In addition, there are varieties of nanoparticles that can be added to the base oils, such as alumina (Al_2O_3), silica (SiO_2), molybdenum sulphide (MoS_2), calcium fluoride (CaF_2), zirconium oxide (ZrO), copper oxide (CuO), zinc oxide (ZnO), etc.

When nanoparticles are used as additives in lubricating oils, they produce four types of lubrication mechanisms: rolling action, tribofilm formation resulting from tribochemical reaction, a self-repairing effect due to minimal size, and polishing effect (refer to Figure 1.3). [27–28].

Cutting Fluids and Lubrication Techniques

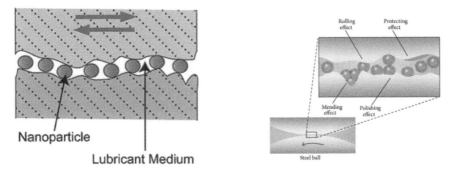

FIGURE 1.3 Lubrication mechanisms in nanofluids [28].

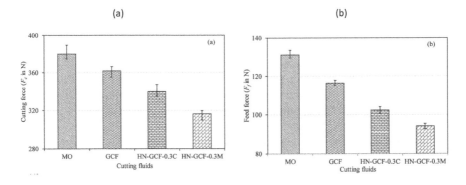

FIGURE 1.4 Influence of cutting fluid application on (a) cutting force and (b) feed force [32].

Besides this, fluid type, nanoparticles type, and concentration of nanoparticles have a powerful influence on the cutting performance. In this regard, the effects of graphite nanoparticles diffused nanofluid on the cutting force and temperature in Minimum Quantity Lubrication (MQL) based lathe turning the process of AISI 1045 sheets of steel were investigated. It was found that nanofluid provided better cutting performance than vegetable oil [29]. Furthermore, a comparative study between multi-walled carbon nanotubes (MWCNT) and aluminium oxide (Al_2O_3) nanoparticles dispersed nanofluid has been performed by Hegab et al. [30] for MQL turning of Inconel 718. The results showed that MWCNT nanofluid gave better cutting performance than Al_2O_3 nanofluid. On similar lines, MoS_2 dispersed soybean oil provided better performance than Al_2O_3 dispersed nanofluid [31].

In MQL machining of hard materials, hybrid nano-green cutting fluid prepared by dispersing MoS_2 and cadmium fluoride (CaF_2) nanoparticles in mineral oil has also been investigated. In this research attempt, green cutting fluid (GCF), indigenously synthesized, was utilized to prepare Hybrid Nano- Green Cutting Fluid (HN-GCF). GCF is manufactured from coconut oil, Cymbopogon citratus, Azadirachtaindica, jaggery syrup, Centella asiatica, and MoS_2 and CaF_2 nanoparticles mixed to prepare nanofluids. The cutting forces and feed force are reduced in HNGCF compared to the neat cutting oils, as shown in Figure 1.4(a) and (b), respectively. HNGCF with

0.3 percent MoS$_2$ nanoparticles have shown the least amount of cutting and feed forces.

1.3 TECHNIQUES OF CLEANER MACHINING

According to the U.S. Department of Commerce, sustainable manufacturing is defined as "the creation of manufactured products using processes that minimize negative environmental impacts, conserve energy and natural resources, are safe for employees, communities, and consumers, and are economically sound" [33]. Thus, it involves the sustainability of all manufactured products and the sustainable manufacturing of all products. The former involves exploring sources associated with sustainable energy, green, and products associated with social equity. The significant facets concerning the latter involve the development and the establishment of efficient processes in energy saving that are economical, non-polluting, and viable for manufacturing purposes (refer to Figure 1.5).

Besides this, the deployment of these methods ensures the well-being of society along with growth in economics through the adaptation of a system-level approach during the complete life-cycle of any product [34–35].

Sustainable machining uses those machining methods, which have a minimal negative impact on the environment and makes less usage of natural resources.

For sustainable machining, the following practices can be adopted:

- Reduce the consumption of natural resources and energy.
- Reduce waste and environmental damage.
- Reduce health and safety risks.
- Provide better performance and economy to the users.

In order to address the problems associated with the machining processes, several sustainable machining methods have been developed over the years. The researchers

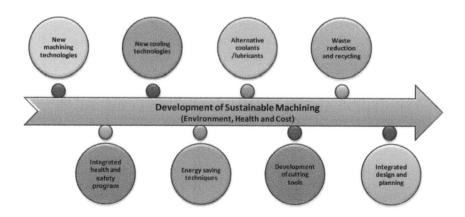

FIGURE 1.5 Road map for developing sustainable green machining technologies [11].

Cutting Fluids and Lubrication Techniques 9

have explored sustainable machining methods like Minimum Quantity Lubrication (MQL), cryogenic, ultrasonically atomized cutting fluids, use of vegetable oils, ionic liquids, and use of nanofluids [34, 36].

The methods of cooling employed during machining processes could be categorized into two parts, i.e., conventional cooling (flood cooling) and non-conventional cooling (cryogenic, dry, minimum quantity cooling, minimum quantity lubrication, and lubrication, etc.) [5, 37].

1.3.1 Conventional Cooling

Flood cooling is a technique in which a high-pressure jet of fluid is provided in the machining zone to produce lubrication, cooling action, and provides evacuation of the chips. This technique mainly aims to reduce the temperature in the machining zone by supplying a high quantity of coolant. It has been the traditional method of cooling which is mainly used in grinding and turning operations where high temperature and sparks are produced. In addition, this process claims to provide a good surface finish and reduction of tool wear [38].

The role of cutting fluids has been investigated using a ceramic tool for machining AISI 4340 steel compared with dry machining. In this study, three different cutting fluid emulsions, synthetic and mineral oils, were used. The experimentation results proved to result in better tool life and low cutting temperature using cutting fluid than dry machining. This is due to the lubrication layer between tool and chip resulting in reduced friction, which eventually improves machining performance [39]. Furthermore, the machining of AISI 52100 steel was performed by using synthetic oil as a cutting fluid. It has been found that cutting force and surface roughness was reduced using flooded cooling compared to dry machining [40]. This is due to the decrease in friction between the tool-workpiece interface, which reduces cutting force and good surface finish.

Conventional flood cooling uses gallons of cutting fluid which leads to an increase in the manufacturing cost. At the same time, there is an issue of disposal of cutting fluid which further increases the cost. Therefore the conventional flood cooling technique is not suitable for sustainable manufacturing, and there is a need to reduce the consumption of cutting fluids in the machining industry. Furthermore, the process provides ample cutting fluid in the cutting zone, and it may create environmental pollution, affect the operator's health, and contribute significantly to the economic aspect of manufacturing [6, 10].

1.3.2 Non-Conventional Cooling

Non-conventional cooling methods aim to reduce the consumption of cutting fluid and address the environmental and health concerns associated with them. The various types of non-conventional cooling methods have been practiced so far, which are explained in the following:

1.3.2.1 Dry Machining

Dry machining is a machining method without the use of any coolant. Dry machining is environmentally safe and desirable for production-based industries in the distant future. Industries are necessitated for looking into green manufacturing due to the enforcement of laws regarding environmental protection for safety during the occupation and health-based regulations. Machining involving no coolant usage, i.e., dry machining, is ideal for reducing environmental and health impacts. Therefore, dry machining has drawn much attention within research groups and manufacturing industries for creating a clean and healthy environment because of the elimination of cutting fluid application [41–42].

Dry machining has numerous advantages involving no threat to the environment, water pollution, or earth pollution, no occupational hazard to health, and no injury to skin or allergy of any sort. Also, the reduction of cost during machining is achieved, and significant cost reduction during maintenance of fluid and disposal. Dry machining is utilized on workpieces ranging from hard materials and materials which are machined with difficulty.

Dry machining requires a harder cutting tool made up of PCD, CBN, etc., and wear-resistant coatings, i.e., TiN and TiCN, etc. Tool coating sustains at high temperature even at large heat generation. Therefore, it does not require cutting fluid for lubrication and cooling purpose. This will lead to a reduction in the tool wear and improve surface integrity. Dry machining is also accompanied by textured inserts with internal cooling systems where solid lubricants are put in the slots of inserts, leading to adequate lubrication [43–44].

In this regard, the effects of WS_2 solid lubricants into a textured cutting insert for the dry machining applications were investigated. For machining hard materials, dry machining cannot be used as it leads to excessive tool wear. In this study, dry cutting of hardened steel was carried out using TiAlN coated tool and a laser textured insert with a solid lubricant filled into the slots to provide self-lubrication. The results of the experimental investigation proved to give better results in cutting force, cutting temperature, and tool wear. This is due to the higher adhesive strength between the WS_2 and textured TiAlN coatings. Besides this, the nano-scale grooves reduced the contact length at the tool/chip interface, which is another reason for improving the tool's cutting performance. In dry machining of ceramics materials, lower thermal conductivity and fracture toughness may lead to early tool fracture due to thermal and mechanical shock. Due to this reason, dry cutting is very difficult to implement in mass production as it needs extremely rigid machine tools and ultra-hard cutting tools [45]. The sustainable machining of AISI 1045 steel with a surface textured tool was investigated. In this study, micro-capillary networks were made on the uncoated cutting inserts for tribological benefits. The investigation proves that machining performance is improved with the textured tools in dry machining [46].

However, despite all the benefits of dry machining, it is not suitable for machining difficult to cut materials like nickel and titanium alloys due to their poor machinability and low thermal conductivity. This eventually results in high thermal stress formation, tensile residual stresses, built-up edge formation, chip segmentation, and self-induced chatter.

1.3.2.2 Minimum Quantity Lubrication (MQL)

Minimum quantity lubrication (MQL) is a micro-lubrication method involving minimal lubricating fluid (10–150 ml/hr.) utilized in the machining zone. This process

Cutting Fluids and Lubrication Techniques

aims to take advantage of both dry and flood cooling techniques. MQL facilitates near to dry machining and removes large amounts of coolants and replaces them with a very minimal amount of cutting fluid mixed with a transporting medium that is generally air. MQL uses a small amount of lubricant that is environment friendly under pressure into the machining zone to reduce the temperature generation by efficient lubrication at the interface of tool and chip. It has lubrication properties similar to oil and cooling capacity similar to carrying medium, replacing conventional flood coolant functions during machining. The air is highly pressurized using a compressor, and cutting fluid already contained in the reservoir passes through the flow control system where the cutting fluid is atomized. This highly pressurized air containing cutting fluid's microdroplets (i.e., aerosol) is sprayed with controlled flow rate values in machining zones utilizing proper tubing and nozzle system [47–50]. Figure 1.6 shows a schematic depiction of the MQL system.

There have been various names used for MQL, like near dry machining, Minimum Quantity Lubrication and Cooling (MQLC), and Minimum Quantity Cooling (MQC). The lubricant's role in MQL is performing the cooling action because of the cutting fluid evaporation rather than lubrication. Even if MQL utilizes cutting fluid, the quantity of coolant used is significantly less compared to flood cooling. Therefore, the cost of the cutting fluid will be reduced, affecting the economics of manufacturing costs. Furthermore, MQL used with ester or vegetable-based lubricants make the process environment-friendly and safe for the workers [5, 51]. Hence, MQL possesses the same benefits as dry machining, involving a neat work environment and less harmful emission levels. Recent research includes nanofluids comprising nanoparticles like Al_2O_3, MoS_2, and diamond and is utilized with MQL systems.

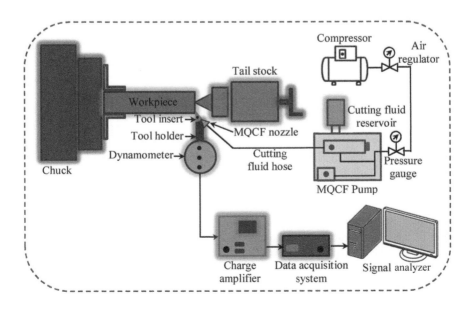

FIGURE 1.6 A schematic illustration of Minimum Quantity Lubrication (MQL) [32].

While drilling A356 Aluminium-Silicon-based alloys, the MQL technique was used, and the results showed a better surface finish for holes drilled using the MQL process than flood cooling [52]. The MQL technique used during the turning process of Inconel 718 with coated tools, and the study revealed that MQL resulted in better life of tool and surface finishing while comparing to dry and wet machining processes [53]. In a similar manner, turning of alloy steel AISI 9310 with MQL technique utilizing vegetable oil-based lubricant is performed. The results showed that the MQL process results in the best surface finishing under the same set of conditions compared to dry and flood lubrication [54]. Further dry, wet, and MQL turning of alloy steel AISI 4140 was investigated and found to show that MQL gave minimum cutting force compared to the other two [55]. It was found that cutting force was reduced by 17.07 percent, whereas tool-tip interfacial temperature by 6.72 percent in MQL turning of AISI 4340 steel compared to dry turning [56].

MQL technique has various benefits like the machined workpiece is almost dry, low installation and running cost, low environmental and health hazards, and better machining performance than traditional flood coolant techniques. However, this technique has few limitations like the lower capability of cooling, evaporation of cutting oil in cutting zone leading to evolving of gases, and requirements of high-pressure values for getting better penetration ability.

1.3.2.3 Cryogenic Cooling

Cryogenic cooling is a safe technique from an environmental point of view. It involves cooling that utilizes gases at cryogenic temperatures like carbon dioxide, nitrogen, and helium as primary cooling agents that evaporate without harming the atmosphere. In the last few years, the techniques associated with cryogenic cooling have drawn much attention, being a potential technique for reducing pollution in the environment, hazards to health-related to traditional cutting fluids, and fulfilling the machining performance requirements. Furthermore, unlike MQL, this technique focuses on keeping the cutting zone temperature to a minimum level [2, 11].

It is sprayed or injected within the cutting zone at around -200°C through a small orifice of the nozzle in machining with liquid nitrogen. Quick absorption of heat along with the evaporation of liquid nitrogen occurs. A layer of fluid protection in gaseous state forms between the tool face and chip acts as a lubricating agent, which reduces the interfacial temperature of the tool-chip and thereby prevents the chemical interactions between chips and cutting tool. This leads to the reduction of adhesive wear at the flank face and avoids diffusion wear. Besides this, the chips produced through cryogenic machining could be recycled in the form of scrap because no residue of oil is now attached to them. These cryogenic gases are generally inert and disperse in the air and pose no danger to the environment and human health. The effect of cooling produced by cryogens is fascinating during machining materials that can't be machined difficult-to-machine materials very quickly and observes very high levels of tool wear because of high temperatures of cutting [57–58]. Figure 1.7 shows a typical cryogenic cooling system.

Even though machining using cryogens is an old technique, a new interest has been generated in this concept for minimizing the hazardous effects of coolants on the environment and humans. The cryogenic cooling utilizing nitrogen in a liquid

Cutting Fluids and Lubrication Techniques 13

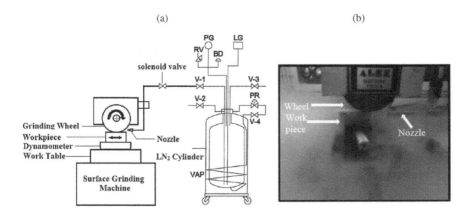

FIGURE 1.7 Schematic diagram of Cryogenic spray System [58].

state could improve machinability by varying the friction coefficient and reducing temperatures in cutting zones. Its advantages involve the prolonged life of the tool, higher speeds of cutting, enhanced productivity, and low cost of production as compared to traditional cutting processes.

The cryogenic machining of steel workpiece shows that a significant improvement in the tool's life, finishing of surface, and accuracy of dimensions associated with the workpiece is obtained compared to wet and dry cooling. Several factors cause this, including reducing the amount of chemical and abrasive wear at tool flanks and minimal tool and chip interactions, thereby decreasing the built-up edge formation [59].

The compressed air suitability for cooling based on the cryogenic application was investigated during the material removal operations in Ti-6Al-4V alloy. The utilization of compressed air in a cryogenic state decreased the wear of the flank and the formation of a built-up chip edge. The result is a minimal enhancement in machining forces (mainly feed force) after machining lengthy distances compared with observations during dry machining [60].

The machinability of Ti-6Al-4V has been assessed under cryogenic cooling conditions [61]. Titanium alloys have broad application areas involving energy, aerospace, and biomedical industries because of elevated strength levels at elevated temperatures, less weight, and corrosion resistance. Still, due to the low thermal conductivity and stiffness, its machinability is extremely poor. The performance of cryogenic turning of titanium alloy has been compared with conventional wet turning. Figure 1.8 shows the effect of cutting speed V_c on surface roughness R_a. It has been found that cryogenic cooling provides significant reduction in surface roughness compared to the wet turning of the machined specimen. Surface roughness is found to increase with the increase of cutting speed.

Furthermore, it has been found that cryogenic turning provides enhanced tool life and reduced Specific Cutting Energy (SCE) than wet turning (refer to Figure 1.9). Reduction in specific energy indicates less power consumption and more Material Removal Rate (MRR). This proves the effectiveness of cryogenic cooling in comparison to conventional wet cooling.

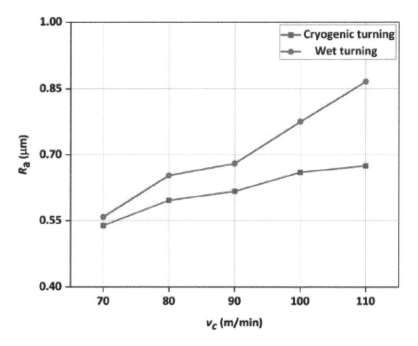

FIGURE 1.8 Variation of R_a with V_c cryogenic and wet turning [61].

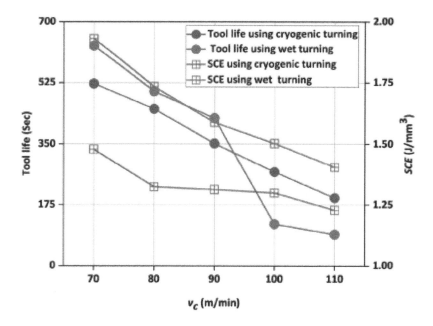

FIGURE 1.9 Tool life and SCE variation at different V_c under cryogenic and wet turning [61].

However, there are several problems associated with cryogenic machining as it can only carry away the heat away from the machining zone but cannot minimize the heat generation. Also, cryogenic technology is a good cooler but does not provide enough lubrication. Nevertheless, the experimental results show that the best results were achieved using cryogenic and MQL techniques. Also, additions of nanofluid have a considerable effect in reducing cutting force and tool wear.

1.3.2.4 Ultrasonic Atomization of Cutting Fluid

Atomization is the process of the formation of tiny droplets from a fluid. There are several techniques of atomization of a fluid, such as pressure atomization, spinning disc atomization, etc. An atomized fluid finds applications in spraying coatings on a substrate, atomized fuel in combustors, drug delivery, analytical nebulizers, etc.

Ultrasonic atomization produces a liquid droplet of the order of 100 μm, usually smaller than those provided by a spray nozzle. This process uses an ultrasonic frequency horn, which produces ultrasonic vibrations which produce an atomized mist of fluid and air [62]. The phenomenon of ultrasonic atomization is based on cavitation theory. In this technique, the liquid is irradiated with ultrasonic waves, generating microbubbles as cavities due to the hydraulic vibration. The cavities so generated collapse due to the expansion and contraction of ultrasonic energy, and fine droplets are produced.

In addition to the mechanism mentioned earlier, capillary wave states that when a liquid film is subjected to ultrasonic vibrations perpendicular to the surface, capillary waves are produced. When the vibration amplitude increases, the wave's amplitude increases, resulting in a taller crest and deeper troughs. At critical amplitude, the wave collapse, and tiny drops of liquids are produced. The so generated droplet size is governed by the capillary wavelength and frequency of the waves [62–63].

The size distribution of droplets produced by ultrasonic atomization depends on ultrasonic frequency. The peak diameter decreases with the increase in ultrasonic frequency. Also, it has been found that an increase in the power intensity of the ultrasonic oscillator increases the surface diameter of the nanosized mist [63]. The effect of ultrasonically atomized cutting fluid application in micro-milling operation is investigated and found to decrease the machining zone temperature, which eventually improves the sustainability of the machining process [64]. Furthermore, the application of vegetable oil in water emulsion by ultrasonic atomization in the micro-milling process is investigated and significantly improves machining forces, less tool wear, thinner chip thickness, and less burr amount [65].

An ultrasonic atomization-based cooling system has been designed and tested for its viability in micro-milling operation. The fluid cutting application consists of an atomizer working on the principle of ultrasonic vibration, a reservoir of cutting fluid, supply of air, and an exhaust based on the vacuum for drawing extra cutting fluid in the form of mist from the air. The atomizer is a piezoelectric-based transducer vibrating at 1 MHz. This cutting fluid used was 5 percent diluted Castrol Clearedge 6519 along with DI water. Experimental results show that low machining forces and improved tool life are obtained when cutting fluid is utilized in an atomized form, opposite to flood cooling and dry cutting. Furthermore, fewer cutting forces were obtained during Castrol 6519 cutting fluid usage than DI water [66].

The ultrasonically atomized droplet can easily penetrate through a very narrow region into the different heat zones of tool work contact and provide adequate lubrication. The droplet characteristics depend on frequency, wavelength, the concentration of additives in fluid, and flow rate.

1.4 CASE STUDY

Sah et al. [67] have developed a novel green cutting fluid and tested its applications in tribological and machining applications. Jatropha oil has been considered the base fluid because of its bio-degradability, high melting point, good thermal and chemical stability. Six types of ionic liquids (ILs; imidazolium based) have been mixed into the base fluid to enhance base oil's physicochemical and thermal properties. In addition, ILs have been added in 1 percent weight in the solution. The details of the ILs have been listed in Table 1.1. The prepared solutions were investigated for their thermophysical properties.

TABLE 1.1

List of Ionic Liquids with Their Molecular Formula and Chemical Structure [67]

Notations	Name of ILs	Purity	Molecular Formula	Chemical Structure
IL1	1-butyl-3-methylimidazolium tetrafluoroborate	98 %	$C_8H_{15}N_2BF_4$	
IL2	1-butyl-3-methylimidazolium hexafluorophosphate	97 %	$C_8H_{15}N_2PF_6$	
IL3	Tetrabutyl ammonium fluoride	99 %	$C_{16}H_{36}NF$	
IL4	Tetrabutyl ammonium bromide	98 %	$C_{16}H_{36}NBr$	
IL5	Tetrabutyl ammonium iodide	99 %	$C_{16}H_{36}NI$	
IL6	Tetrabutyl ammonium hexafluorophosphate	98 %	$C_{16}H_{36}NPF_6$	

Cutting Fluids and Lubrication Techniques

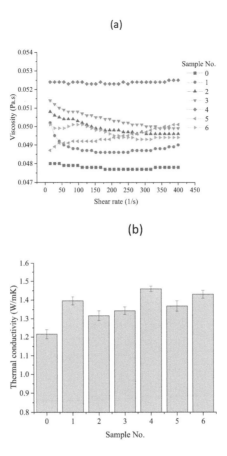

FIGURE 1.10 (a) Viscosity study of ILs solutions (b) Thermal conductivity study of ILs solutions [67].

Figure 1.10 represents the viscosity and thermal conductivity graphs for the prepared ILs solutions. The addition of ILs enhances viscosity and thermal conductivity, which is required for a lubricant. IL4 has shown the maximum improvement in viscosity and thermal conductivity. The ionic liquid solution that has enhanced the viscosity and thermal conductivity highest of base oil has been further used to develop nanofluids. Further, Al_2O_3 and ZrO_2 nanoparticles have been used as nano-additives to increase the base fluid's lubricating and load-carrying capacity. Nanoparticles have been added in 0.5 percent concentration in the solution. The cutting fluids were developed in various combinations, which are listed in Table 1.2.

The prepared nanofluids were further investigated for their properties and characterization, which are explained in the following:

1.4.1 Stability Study

The stability of a nanofluid is a prime challenge for researchers as nanoparticles have a tendency to settle down in the solution. Therefore, nanoparticles must stay

TABLE 1.2
Sample Details of Prepared Nanofluids [67]

Sample	Sample details
JO	Jatropha oil only
JO+IL4	Jatropha oil mixed with 1 % ionic liquid 4 (Tetrabutyl ammonium bromide)
JO+IL4+0.5 % Al_2O_3	Jatropha oil with 1 % ionic liquid 4 and 0.5 % alumina nanoparticles
JO+IL4+0.5 % ZrO_2	Jatropha oil with 1 % ionic liquid 4 and 0.5 % zirconia nanoparticles
JO+IL4+0.25 % Al_2O_3+ 0.25 % ZrO_2	Jatropha oil with 1 % Ionic liquid 4 and 0.25 % alumina and 0.25 % zirconia nanoparticles

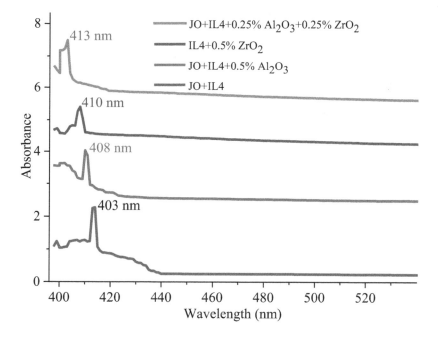

FIGURE 1.11 Stability study of nanofluids [67].

in the state of Brownian motion in the solution in order to sustain their stability. In the present study, the stability of nanofluid has been investigated using Ultraviolet (UV)-visible spectroscopy. UV-vis technique measures the level of absorbance of a particular wavelength when light is allowed to pass through it. Figure 1.11 shows the degree of absorbance as a function of wavelength. It is clear that hybrid nanofluids have more stability than unitary nanofluids.

1.4.2 Characterization of Nanofluids

The prepared nanofluids were investigated for their characteristics like thermal conductivity, viscosity, and spreadability. Since these properties play a significant role for a cutting fluid to be used as a lubricant, a lubricant must have adequate viscosity so that it can create a thick layer between contact surfaces during machining. Figure 1.12 shows the effect of nanoparticles on the viscosity of nanofluids as a function of shear rate. As the shear rate increases, the viscosity of the nanofluids decreases, which shows the shear-thinning behaviour of the nanofluids. Furthermore, with the addition of nanoparticles, viscosity is found to increase, showing the effectiveness of nanoparticles. Hybrid nanofluid containing both Al_2O_3 and ZrO_2 nanoparticles shows the maximum improvement in the viscosity compared to unitary nanofluids.

The thermal conductivity of a cutting fluid is an essential property required in machining operations as a lot of heat is generated. In order to have an effective heat transfer, cutting fluid must have good thermal conductivity. The thermal conductivity of the prepared nanofluids was measured. Figure 1.13 shows the effect of nanoparticles on the thermal conductivity of nanofluids at different temperatures. It is evident that nanoparticles improve the thermal conductivity of nanofluids. Hybrid nanofluid shows the highest thermal conductivity amongst the prepared lubricants. Also, thermal conductivity was found to increase with increasing temperature, which is a must for a lubricant.

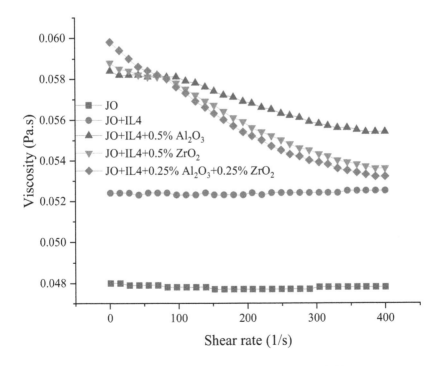

FIGURE 1.12 Effect of nanoparticles on the viscosity of the base fluid.

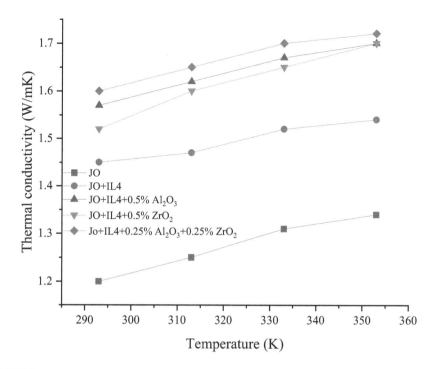

FIGURE 1.13 Effect of temperature on thermal conductivity of nanofluids [67].

When a liquid falls on the solid surface, it must spread well in order to play a lubricating role. Spreadability is the ability of a liquid to spread on the surface. It is an essential property required for lubricants. The spreading coefficient indicates the measurement of spreadability. It is the algebraic difference between the adhesive forces of solid-liquid interfaces and cohesive forces between liquid molecules. A spreading coefficient value more than zero indicates total spreading, and a value less than zero indicates the partial spreading nature of the liquid. Figure 1.14 depicts the spreading coefficient of all prepared nanofluids. All the fluids have a negative spreading coefficient that deduces the partial spreading nature of fluids. Hybrid nanofluid shows a spreading coefficient closer to zero amongst prepared fluids shows the highest chances of spreading.

1.4.3 Tribological Study

Cutting fluids must have good tribological properties in order to provide adequate lubrication. In the present study, the characteristic tribological investigation of cutting fluids had been performed through a pin-on-disc test. Nickel-based Nimonic 80 alloy (Hardness 45 HRC) has been utilized as a disk (100 mm diameter, Ra ≈0.03μm) and a tungsten carbide pin (cylindrical shape, 8 mm diameter) with a hardness of 85 HRC used for the contact pairs. After conducting the pin-on-disc test, the disc's

Cutting Fluids and Lubrication Techniques   21

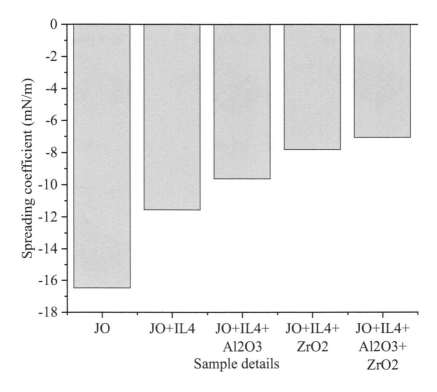

FIGURE 1.14 Spreading coefficient of nanofluids [67].

worn-out surfaces have been analyzed using a 3D microscope. Figure 1.15 depicts the topography of surfaces used under various lubricating conditions.

The properties of lubricants play a vital role in reducing friction at the contact surfaces. Dry friction results in higher irregularities on the surface, having an average surface roughness of 0.378 μm. It is due to the abrasive wear of the surface resulting in catastrophic failure. Cutting fluid application has improved surface topography and average surface roughness due to the development of a lubricating film that prevents pin to disc contact. The nanoparticle addition fills up the microcracks and asperities, which further improves the surface topography.

1.4.4 Machining Study

A round bar of Super Duplex Stainless Steel (SDSS) with a diameter of 60 mm and 300 mm length was used for the machining experiment. Based on experimental results for the characterization of the cutting fluids, IL4 shown the best lubricant additive into JO out of all six ILs considered in this study. Therefore, IL4 is selected as a lubricant additive into JO for the machining performance on SDSS. Also, Al_2O_3, ZrO_2, and $Al_2O_3+ZrO_2$ NPs are added to the JO+IL4 solution and used in the machining experiment under MQL conditions. The hardness of the workpiece is found as

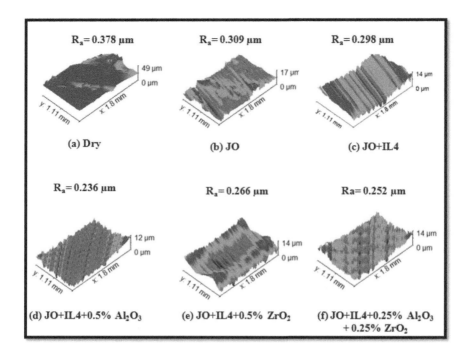

FIGURE 1.15 3D micrograph of worn surfaces with nanofluids [67].

TABLE 1.3
Cutting Fluid Descriptions

Cutting environment	Cutting fluid
Dry	No cutting fluid
MQL 1	Neat JO
MQL 2	JO + 1 wt.% IL4
MQL 3	JO + 1 wt.% IL4 + 0.5 wt.% Al_2O_3
MQL 4	JO + 1 wt.% IL4 + 0.5 wt.% ZrO_2
MQL 5	JO + 1 wt.% IL4 + 0.25 wt.% Al_2O_3 + 0.25 wt.% ZrO_2

51.4 HRC, measured by microhardness tester using the load of 5 N. The following sections describe the results for machining of SDSS under dry and MQL conditions.

The turning experiment was performed on an NH22 lathe machine (Make: HMT India). The coated carbide inserts, which have a designation of CNMG 120408-LM and the tool holder with a specification of DCLNR2525M12, have been used in the current work. The cutting parameters used for carrying out the turning experiments were cutting speed as 80 m/min, feed rate as 0.12 m/rev. And the depth of cut has been taken as 0.3 mm. In addition, the MQL setup parameters were taken as Air pressure=3 bar, flow rate= 120 mm/hr, Standoff distance=10 mm, and nozzle inclination angle was kept as 45°. Table 1.3 shows the cutting fluids used in the machining experiments.

Cutting Fluids and Lubrication Techniques

Cutting force (Fz), feed force (Fy), and thrust force (Fx) have been measured during the turning of SDSS using five different cutting fluids under MQL conditions and also in a dry condition as shown in Figure 1.16. The cutting forces have been measured by the Kistler dynamometer (three-component piezoelectric type) installed on the lathe. The cutting force results are recorded and analyzed in Kistler DynoWare software. The result indicates that the cutting force is decreased by 2.24 percent for JO and 8.54 percent for JO+IL4 under MQL conditions compared to dry machining. ZrO_2 NPs as an additive into JO+IL4 show better results than Al_2O_3 and $Al_2O_3+ZrO_2$ NPs, which shows a 15.97 percent decrement in cutting force compared to dry machining.

This case study shows the effectiveness of a plant-based JO mixed with ILS and nanoparticles in tribological and machining applications. The prepared nanofluids in different combinations can replace the conventional petroleum-based lubricants in the machining industry, providing a sustainable solution.

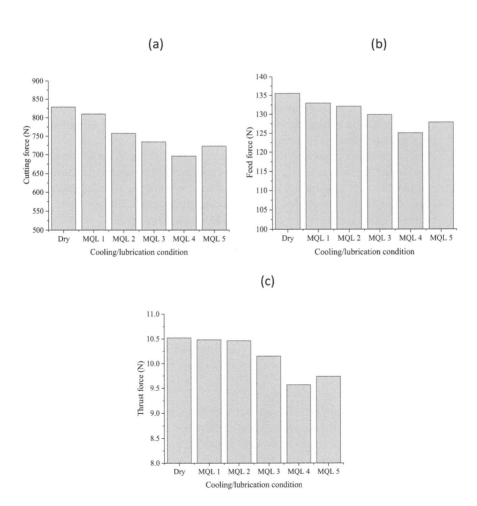

FIGURE 1.16 Cutting forces under different cooling conditions.

1.5 CONCLUSIONS

A sustainable lubrication system helps in reducing the machining cost and improves the machining performance. The present review work has provided an overview of cutting fluids used in the machining industry and the progression of techniques of lubrication and associated challenges. The key findings of the work have been listed here:

- Cutting fluids are essential in the machining of rigid materials. However, conventional coolants and traditional lubrication approaches result in high machining costs, an unsafe environment for workers, and a threat to the ecological system.
- Vegetable oils possess excellent lubricating properties compared to mineral-based oils and lubricants, which are synthetic, along with addressing the ecological aspect in machining. In addition, they have been found to give a good surface finish, low cutting forces, and provides a cleaner cutting environment.
- Nanoparticles dispersed in the base oil helps to minimize the friction by changing the mechanisms of lubrication, resulting in lower cutting forces, less tool wear, and improve surface topography. In addition, the mixing of additives has shown promising results in enhancing the stability of the emulsions and improved machining performance. For research in the future, there should be enhanced focus over the application of novel lubricants in machining materials that can't be easily machined.
- The application of cutting fluid also plays a crucial role in addressing sustainability in the machining processes. Dry machining, which is considered an ideal case of machining, is only feasible at low cutting speeds. However, excessive heat is generated at high cutting speed, resulting in decreased surface quality, excessive tool wear, and chip disposal issues.
- MQL is one technique in which minimal coolant amounts are being supplied at a very high speed into the machining zone, reaching the narrow heating zones to provide adequate lubrication. MQL has shown better machining performances in the various research attempts that have been made. However, MQL provides good lubrication but lacks cooling action.
- Cryogenic machining is a technique involving the temperature of the cutting zone is kept at very low with liquid nitrogen or some other gas. The cryogenic process gives effective cooling and reduces thermal stresses. However, it lacks lubrication and results in poor surface finish and fumes in the machining zone.

From the literature, it can be inferred that significant research work has been done to develop coolants and their application technology and systems from a machining performance perspective. However, future research needs to be done from an ecological and human perspective. There is a need to develop the ecological assessment of coolants and their application techniques. Also, the measurement of toxicity of cutting fluids is still a challenging task that has enormous potential for researchers.

REFERENCES

[1] A. Shokrani, V. Dhokia, and S. T. Newman, "Environmentally conscious machining of difficult-to-machine materials with regard to cutting fluids," *Int. J. Mach. Tools Manuf.*, vol. 57. pp. 83–101, 2012, doi: 10.1016/j.ijmachtools.2012.02.002.

[2] K. Gupta and R. F. Laubscher, "Sustainable machining of titanium alloys: A critical review," *Proc. Inst. Mech. Eng. Part B J. Eng. Manuf.*, vol. 231, no. 14, pp. 2543–2560, 2017, doi: 10.1177/0954405416634278.

[3] J. Kundrák, A. G. Mamalis, K. Gyáni, and A. Markopoulos, "Environmentally friendly precision machining," *Mater. Manuf. Process.*, vol. 21, no. 1, pp. 29–37, 2006, doi: 10.1080/AMP-200060612.

[4] V. Sharma and P. M. Pandey, "Recent advances in ultrasonic assisted turning: A step towards sustainability," *Cogent Eng.*, vol. 3, no. 1, pp. 1–20, 2016, doi: 10.1080/23311916.2016.1222776.

[5] P. J. Liew, A. Shaaroni, N. A. C. Sidik, and J. Yan, "An overview of current status of cutting fluids and cooling techniques of turning hard steel," *Int. J. Heat Mass Transf.*, vol. 114. pp. 380–394, 2017, doi: 10.1016/j.ijheatmasstransfer.2017.06.077.

[6] M. Osama, A. Singh, R. Walvekar, M. Khalid, T. C. S. M. Gupta, and W. W. Yin, "Recent developments and performance review of metal working fluids," *Tribol. Int.*, vol. 114, pp. 389–401, 2017, doi: 10.1016/j.triboint.2017.04.050.

[7] A. Anand, K. Vohra, M.I.U. Haq, A. Raina, and M.F. Wani, "Tribology in industry tribological considerations of cutting fluids in machining environment: A review," *Tribol. Ind.*, vol. 38, pp. 463–474, 2016.

[8] S. Jahanmir, "Tribology issues in machining," *Mach. Sci. Technol.*, vol. 2, no. 1, pp. 137–154, 1998, doi: 10.1080/10940349808945663.

[9] Niosh, "What you need to know about occupational exposure to metalworking fluids," DHHS (NIOSH) Pub 98–116, 1998.

[10] NIOSH, "Metalworking fluid exposure at an aircraft engine manufacturing," Niosh, no. August, 2012.

[11] J. Haider and M. S. J. *Hashmi, Health and Environmental Impacts in Metal Machining Processes*, vol. 8. Elsevier, 2014.

[12] C. Tool, M. Tool, S. Roughness, and W. Grzesik, "Learn more about cutting fluid cutting fluids," in *Advanced Machining Processes of Metallic Materials (Second Edition), Theory, Modelling, and Applications*, 2017, pp. 183–195.

[13] Y. M. Shashidhara and S. R. Jayaram, "Vegetable oils as a potential cutting fluidan evolution," *Tribol. Int.*, vol. 43, no. 5–6, pp. 1073–1081, 2010, doi: 10.1016/j.triboint.2009.12.065.

[14] S. A. Lawal, I. A. Choudhury, and Y. Nukman, "Application of vegetable oil-based metalworking fluids in machining ferrous metals—A review," *Int. J. Mach. Tools Manuf.*, vol. 52, no. 1, pp. 1–12, 2012, doi: 10.1016/j.ijmachtools.2011.09.003.

[15] V. K. A. V, N. S. S, and C. Ramprasad, "Vegetable oil-based metal working fluids-a review vegetable oil-based metal working fluids-a review," *Int. J. Theor. Appl. Res. Mech. Eng.*, vol. 1, pp. 95–101, 2012.

[16] E. Kuram, B. Ozcelik, M. Bayramoglu, E. Demirbas, and B. T. Simsek, "Optimization of cutting fluids and cutting parameters during end milling by using D-optimal design of experiments," *J. Clean. Prod.*, vol. 42, pp. 159–166, 2013, doi: 10.1016/j.jclepro.2012.11.003.

[17] H. H. Masjuki, M. A. Maleque, A. Kubo, and T. Nonaka, "Palm oil and mineral oil based lubricants—their tribological and emission performance," *Tribol. Int.*, vol. 32, no. 6, pp. 305–314, 1999, doi: 10.1016/S0301-679X(99)00052-3.

[18] N. H. Jayadas, K. Prabhakaran Nair, and A. G, "Tribological evaluation of coconut oil as an environment-friendly lubricant," *Tribol. Int.*, vol. 40, no. 2 SPEC. ISS., pp. 350–354, 2007, doi: 10.1016/j.triboint.2005.09.021.

[19] A. Ruggiero, R. D'Amato, M. Merola, P. Valašek, and M. Müller, "Tribological characterization of vegetal lubricants: Comparative experimental investigation on *Jatropha curcas* L. oil, Rapeseed Methyl Ester oil, Hydrotreated Rapeseed oil," *Tribol. Int.*, vol. 109, no. January, pp. 529–540, 2017, doi: 10.1016/j.triboint.2017.01.030.

[20] M. Chanes de Souza, J. Fracaro de Souza Gonçalves, P. Cézar Gonçalves, S. Yuji Sudo Lutif, and J. de Oliveira Gomes, "Use of Jatropha and Moringa oils for lubricants: Metalworking fluids more environmental-friendly," *Ind. Crops. Prod.*, vol. 129. pp. 594–603, 2019, doi: 10.1016/j.indcrop.2018.12.033.

[21] M. Susmitha, P. Sharan, and P. N. Jyothi, "Influence of non-edible vegetable based oil as cutting fluid on chip, surface roughness and cutting force during drilling operation of Mild Steel," *IOP Conf. Ser. Mater. Sci. Eng.*, vol. 149, no. 1, 2016, doi: 10.1088/1757-899X/149/1/012037.

[22] M. A. Mahadi, I. A. Choudhury, M. Azuddin, N. Yusoff, A. A. Yazid, and A. Norhafizan, "Vegetable oil-based lubrication in machining: Issues and challenges," *IOP Conf. Ser. Mater. Sci. Eng.*, vol. 530, no. 1, 2019, doi: 10.1088/1757-899X/530/1/012003.

[23] W. Yu and H. Xie, "A review on nanofluids: Preparation, stability mechanisms, and applications," *J. Nanomater.*, vol. 2012, 2012, doi: 10.1155/2012/435873.

[24] A. K. Sharma, A. K. Tiwari, and A. R. Dixit, "Progress of nanofluid application in machining: A review," *Mater. Manuf. Process.*, vol. 30, no. 7, pp. 813–828, 2015, doi: 10.1080/10426914.2014.973583.

[25] S. Khandekar, M. R. Sankar, V. Agnihotri, and J. Ramkumar, "Nano-cutting fluid for enhancement of metal cutting performance," *Mater. Manuf. Process.*, vol. 27, no. 9, pp. 963–967, 2012, doi: 10.1080/10426914.2011.610078.

[26] S.S. Chaudhari, R.R. Chakule, and P. S. Talmale, "Experimental study of heat transfer characteristics of Al2O3 and CuO nanofluids for machining application," *Mater. Today: Proc.,* vol. 18, pp. 788–797, 2019.

[27] L. Kong, J. Sun, and Y. Bao, "Preparation, characterization and tribological mechanism of nanofluids," *RSC Adv.*, vol. 7, no. 21, pp. 12599–12609, 2017, doi: 10.1039/c6ra28243a.

[28] I. Ali, A.A. Basheer, A. Kucherova, N. Memetov, T. Pasko, K. Ovchinnikov, V. Pershin, D. Kuznetsov, E. Galunin, V. Grachev, and A. Tkachev, "Advances in carbon nanomaterials as lubricants modifiers," *J. Mol. Liq.*, vol. 279, pp. 251–266, 2019. https://doi.org/10.1016/j.molliq.2019.01.113.

[29] Y. Su, L. Gong, B. Li, Z. Liu, and D. Chen, "Performance evaluation of nanofluid MQL with vegetable-based oil and ester oil as base fluids in turning," *Int. J. Adv. Manuf. Technol.*, vol. 83, no. 9–12, pp. 2083–2089, 2016.

[30] H. Hegab, U. Umer, M. Soliman, and H. A. Kishawy, "Effects of nano-cutting fluids on tool performance and chip morphology during machining Inconel 718," *Int. J. Adv. Manuf. Technol.*, vol. 96, no. 9–12, pp. 3449–3458, 2018, doi: 10.1007/s00170-018-1825-0.

[31] T. T. Long, T. Q. Chien, and Tran Minh Duc, "Performance evaluation of MQL parameters using," *Lubricants*, vol. 7, no. 5, p. 40, 2019.

[32] K. K. Gajrani, P. S. Suvin, S. V. Kailas, and R. S. Mamilla, "Thermal, rheological, wettability and hard machining performance of MoS2 and CaF2 based minimum quantity hybrid nano-green cutting fluids," *J. Mater. Process. Technol.*, vol. 266, pp. 125–139, 2019, doi: 10.1016/j.jmatprotec.2018.10.036.

[33] K. R. Haapala et al., "A review of engineering research in sustainable manufacturing," *J. Manuf. Sci. Eng. Trans. ASME*, vol. 135, no. 4, pp. 1–16, 2013, doi: 10.1115/1.4024040.

[34] S. A. Lawal, I. A. Choudhury, and Y. Nukman, "A critical assessment of lubrication techniques in machining processes: A case for minimum quantity lubrication using vegetable oil-based lubricant," *J. Clean. Prod.*, vol. 41, pp. 210–221, 2013, doi: 10.1016/j.jclepro.2012.10.016.

[35] M. S. Najiha, M. M. Rahman, and A. R. Yusoff, "Environmental impacts and hazards associated with metal working fluids and recent advances in the sustainable systems: A review," *Renew. Sustain. Energy Rev.*, vol. 60, pp. 1008–1031, 2016, doi: 10.1016/j. rser.2016.01.065.

[36] R. M'Saoubi et al., "High performance cutting of advanced aerospace alloys and composite materials," *CIRP Ann.—Manuf. Technol.*, vol. 64, no. 2, pp. 557–580, 2015, doi: 10.1016/j.cirp.2015.05.002.

[37] Chetan, S. Ghosh, and P. Venkateswara Rao, "Application of sustainable techniques in metal cutting for enhanced machinability: A review," *J. Clean. Prod.*, vol. 100, pp. 17–34, 2015, doi: 10.1016/j.jclepro.2015.03.039.

[38] R. R. Mishra, R. Kumar, A. K. Sahoo, and A. Panda, "Machinability behaviour of biocompatible Ti-6Al-4V ELI titanium alloy under flood cooling environment," *Mater. Today Proc.*, no. xxxx, 2019, doi: 10.1016/j.matpr.2019.05.402.

[39] R.F. Avila, and A.M. Abrao, "The effect of cutting ūids onthe machining of hardened AISI 4340 steel," *J. Mater. Process. Technol.*, vol. 119, no. 3, pp. 21–26, 2001, doi: 10.1016/S0924-0136(01)00891-3.

[40] M. Biček, F. Dumont, C. Courbon, F. Pušavec, J. Rech, and J. Kopač, "Cryogenic machining as an alternative turning process of normalized and hardened AISI 52100 bearing steel," *J. Mater. Process. Technol.*, vol. 212, no. 12, pp. 2609–2618, 2012, doi: 10.1016/j.jmatprotec.2012.07.022.

[41] P. S. Sreejith and B. K. A. Ngoi, "Dry machining: Machining of the future," *J. Mater. Process. Technol.*, vol. 101, no. 1, pp. 287–291, 2000, doi: 10.1016/S0924-0136(00)00445-3.

[42] G. S. Goindi and P. Sarkar, "Dry machining: A step towards sustainable machining—Challenges and future directions," *J. Clean. Prod.*, vol. 165, pp. 1557–1571, 2017, doi: 10.1016/j.jclepro.2017.07.235.

[43] K. Weinert, I. Inasaki, J. W. Sutherland, and T. Wakabayashi, "Dry machining and minimum quantity lubrication," *CIRP Ann.—Manuf. Technol.*, vol. 53, no. 2, pp. 511–537, 2004, doi: 10.1016/S0007-8506(07)60027-4.

[44] F. Klocke and G. Eisenblätter, "Dry cutting—State of research," *VDI Berichte*, vol. 46, no. 1399, pp. 159–188, 1998.

[45] K. Zhang, J. Deng, Z. Ding, X. Guo, and L. Sun, "Improving dry machining performance of TiAlN hard-coated tools through combined technology of femtosecond laser-textures and WS2 soft-coatings," *J. Manuf. Process.*, vol. 30, pp. 492–501, 2017, doi: 10.1016/j.jmapro.2017.10.018.

[46] S. Dhage, A. D. Jayal, and P. Sarkar, "Effects of surface texture parameters of cutting tools on friction conditions at tool-chip interface during dry machining of AISI 1045 steel," *Procedia Manuf.*, vol. 33, pp. 794–801, 2019, doi: 10.1016/j.promfg.2019.04.100.

[47] A. Kumar Sharma, A. Kumar Tiwari, A. Rai Dixit, and R. Kumar Singh, "Measurement of machining forces and surface roughness in turning of AISI 304 steel using alumina-MWCNT hybrid nanoparticles enriched cutting fluid," *Meas. J. Int. Meas. Confed.*, vol. 150, p. 107078, 2020, doi: 10.1016/j.measurement.2019.107078.

[48] G. S. Goindi, S. N. Chavan, D. Mandal, P. Sarkar, and A. D. Jayal, "Investigation of ionic liquids as novel metalworking fluids during minimum quantity lubrication machining of a plain carbon steel," *Procedia CIRP*, vol. 26, pp. 341–345, 2015, doi: 10.1016/j. procir.2014.09.002.

[49] B. Sen, M. Mia, G.M.K. Uttam, K. Mandal, and S. Prasad, "Eco—Friendly cutting fluids in minimum quantity lubrication assisted machining: A review on the perception of sustainable manufacturing," 2021. https://doi.org/10.1007/s40684-019-00158-6.

[50] E. García-martínez, V. Miguel, A. Martínez-martínez, M.C. Manjabacas, J. Coello, "Sustainable lubrication methods for the machining of titanium alloys: An overview," pp. 1–22, 2019. https://doi.org/10.3390/ma12233852.

[51] H. A. Hegab, B. Darras, and H. A. Kishawy, "Towards sustainability assessment of machining processes," *J. Clean. Prod.*, vol. 170, pp. 694–703, 2018, doi: 10.1016/j.jclepro.2017.09.197.

[52] D.U. Braga, A. E. Dhiniz, G. W. A. Miranda, and N. L. Coppini, "Using a minimumquantity of lubricant and diamond coated tool in the drilling of aluminiumsiliconalloys," *J. Mater. Process. Technol.*, vol. 122, pp. 127–138, 2002.

[53] Y. Kamata and T. Obikawa, "High speed MQL finish-turning of Inconel 718 withdifferent coated tools," *J. Mater. Process. Technol.*, vol. 192–193, pp. 281–286, 2007.

[54] M. M. A. Khan, M. A. H. Mithu, and N. R. Dhar, "Effect of minimum quantitylubrication on turning AISI 9310 alloy steel using vegetable oil-based cutting," *J. Mater. Process. Technol.*, vol. 209, pp. 5573–5583, 2009.

[55] M. Hadad and B. Sadeghi, "Minimum quantity lubrication-MQL turning of AISI 4140 steel alloy," *J. Clean. Prod.*, vol. 54, pp. 332–343, 2013, doi: 10.1016/j.jclepro.2013.05.011.

[56] A. Saini, S. Dhiman, R. Sharma, and S. Setia, "Experimental estimation and optimization of process parameters under minimum quantity lubrication and dry turning of AISI-4340 with different carbide inserts," *J. Mech. Sci. Technol.*, vol. 28, no. 6, pp. 2307–2318, 2014, doi: 10.1007/s12206-014-0521-8.

[57] E. Benedicto, D. Carou, and E. M. Rubio, "Technical, economic and environmental review of the lubrication/cooling systems used in machining processes," *Procedia Eng.*, vol. 184, pp. 99–116, 2017, doi: 10.1016/j.proeng.2017.04.075.

[58] P. P. Reddy and A. Ghosh, "Some critical issues in cryo-grinding by a vitrified bonded alumina wheel using liquid nitrogen jet," *J. Mater. Process. Technol.*, vol. 229, pp. 329–337, 2016, doi: 10.1016/j.jmatprotec.2015.09.040.

[59] N. R. Dhar, S. Paul, and A. B. Chattopadhyay, "Machining of AISI 4340 steel under cryogenic cooling-tool wear, surface roughness and dimensional deviation," *J. Mater. Process. Technol.*, vol. 123, no. 3, pp. 483–489, 2002, doi: 10.1016/S0924-0136(02)00134-6.

[60] S. Sun, M. Brandt, and M. S. Dargusch, "Machining Ti-6Al-4V alloy with cryogenic compressed air cooling," *Int. J. Mach. Tools Manuf.*, vol. 50, no. 11, pp. 933–942, 2010, doi: 10.1016/j.ijmachtools.2010.08.003.

[61] C. Agrawal, J. Wadhwa, A. Pitroda, C. I. Pruncu, M. Sarikaya, and N. Khanna, "Comprehensive analysis of tool wear, tool life, surface roughness, costing and carbon emissions in turning Ti—6Al—4V titanium alloy: Cryogenic versus wet machining," *Tribol. Int.*, vol. 153, August 2020, 2021, doi: 10.1016/j.triboint.2020.106597.

[62] N. R. Sikwal et al., "Ultrasound-assisted preparation of ZnO nanostructures: Understanding the effect of operating parameters," *Green Process. Synth.*, vol. 5, no. 2, pp. 163–172, 2016, doi: 10.1515/gps-2015-0072.

[63] T. Kudo, K. Sekiguchi, K. Sankoda, N. Namiki, and S. Nii, "Effect of ultrasonic frequency on size distributions of nanosized mist generated by ultrasonic atomization," *Ultrason. Sonochem.*, vol. 37, pp. 16–22, 2017, doi: 10.1016/j.ultsonch.2016.12.019.

[64] M. Rukosuyev, C. S. Goo, and M. B. G. Jun, "Understanding the effects of the system parameters of an ultrasonic cutting fluid application system for micro-machining," *J. Manuf. Process.*, vol. 12, no. 2, pp. 92–98, 2010, doi: 10.1016/j.jmapro.2010.06.002.

[65] G. Burton, C. Goo, Y. Zhang, and M. B. G. Jun, "Use of vegetable oil in water emulsion achieved through ultrasonic atomization as cutting fluids in micro-milling," *J. Manuf. Process.*, vol. 16, pp. 405–413, 2014.

[66] M. B. G. Jun and R. E. Devor, "An experimental evaluation of an atomization-based cutting fluid," vol. 130, June 2008, pp. 1–8, 2019, doi: 10.1115/1.2738961.

[67] N. Kumar, R. Singh, and V. Sharma, "Experimental investigations into thermophysical, wettability and tribological characteristics of ionic liquid based metal cutting fluids," *J. Manuf. Process.*, vol. 65, pp. 190–205, 2021. https://doi.org/10.1016/j.jmapro.2021.03.019.

2 Electrical Discharge Drilling of Al-SiC Composite with Gas-Assisted Multi-Hole Rotary Slotted Tool

Nishant K. Singh, Yashvir Singh,
Bholey Singh, and Hardik Berwal

2.1	Introduction	29
2.2	Materials and Methods	31
	2.2.1 Details of Electrode Material	31
	2.2.2 Tool Design	31
	2.2.3 Experimental Procedure	32
2.3	Data Analysis on the Experimental Observations	33
	2.3.1 Response Surface of MRR	37
	2.3.2 Response Surface of EWR	39
2.4	Results and Discussion	42
	2.4.1 The Influence of Control Factors on MRR	42
	2.4.2 Effect of Process Parameters on EWR	44
	2.4.3 Surface Morphology	46
	2.4.4 Process Optimization and Accuracy of Models	47
	2.4.5 Genetic Algorithm-Based Multi-Objective Optimization	47
2.5	Conclusions	48
References		49

2.1 INTRODUCTION

Modern ceramics materials are appealing for numerous applications because of their superior mechanical properties and lightweight. Moreover, the application of ceramic composites have not been predominantly applied because of too many drawbacks, like low toughness and the high cost of fabrication. Electrical Discharge Machining (EDM) is a popular technology for producing chemical, aviation, and automobile parts [1]. Because of the presence of a reinforcing agent, EDM machining of composite materials is difficult and time consuming. Despite composite better qualities, conducting ceramic composite EDM has acquired considerable industry

DOI: 10.1201/9781003220237-2

adoption [2]. These are some of the major challenges faced during EDM machining of the ceramic composite is the flushing of fragments from the electrodes gap. The formation of electric arcs and short circuiting take place when debris collected in the electrodes passage, lowering the material removal rate (MRR) and surface sturdiness [3–4].

Various unorthodox approaches have been used with EDM to improve its machinability. A lot of research shows that substantial EDM performances are significantly governed by both electrical and non-electrical parameters [5–8]. Studies indicated that MRR improved at the commencement of the machining operation and then began to decline due to blockage of the debris in the region between tool and workpiece [9–13]. To respond with the restructuring of the systems integration or a modification in tool configuration, investigators have utilized several unconventional ways. Non-electrical characteristics like discharge gas pressure and tool speed are crucial in enhancing operational efficiency [14–16]. Very few findings [17–19] explored compressed gas supply to promote EDM performance. The supply of compressed gas promotes the flushing of material from the spark area, which raises MRR. Additionally, experimental research suggests that using several creative [20–21] flushing techniques results in a better surface quality and less drilling time.

According to the existing literature, spark erosion machining of ceramic composites has proved successful. However, the most critical concerns that developed during the machining of MMCs were the congestion of non-conducting ceramic particles in electrodes spaces, which impeded the EDM process's spark generation. Detailed literature study, it can be concluded that better flushing performance can be used to enhance the productivity of EDM operations many researchers have proposed different techniques, such as, rotary EDM [3–5], and EDM with different tool geometry [14–17]. Apart from that, little study has been conducted in the literature to use the advantages of compressed gas flow in the standard die sinking EDM method. Singh et al. [17, 22] discovered the benefit of gas-assisted EDM over rotational EDM in terms of MRR, EWR, and SR in previous experimental studies. Hybrid machining process (HMP) is the incorporation of at least two machining methods. HMP is utilized to machine rigid and hardened materials that can't be machined adequately by an independent machining method. HMP is utilized to extricate the advantages of every individual procedure together and to enhance procedure stability and effectiveness. The present research work emphasized on advancing a HMP procedure utilizing both fluid and gas as dielectrics, to extricate the advantage of gas based EDM and conventional oil EDM while machining of Al-SiC composite during EDM operation. Reduced machining costs must be used to encourage the usage of the EDM technique. As a result, optimization approaches are better ways to save time and expense for the examination of the proposed model in order to reduce machining costs.

In this study, compressed air is introduced into the discharge gap via a multi-hole slotted rotary tool to investigate the effect of air flow on EDM performance during drilling operations. The investigations were executed to evaluate the MRR and EWR of the AAEDD and REDD processes. The goal of this research is to look into the statistical modeling and optimization of MRR and EWR using pressurized air via a multi-hole slotted rotary electrode in conventional EDD, which has not been reported in the literature to date.

The task in this paper is divided into two stages. During the AAEDD process, the first phase focused on developing statistically established MRR and EWR models. To analyze the influence of applying pressurized air via the multi-hole slotted rotary electrode in traditional EDD, the main effect and interactions among processing parameters on MRR and EWR were assessed. The second part is a comparative analysis of the REDD and AAEDD processes in terms of MRR and EWR.

2.2 MATERIALS AND METHODS

2.2.1 DETAILS OF ELECTRODE MATERIAL

The experimentation was carried out with Al-SiC composite as the work sample and copper as the electrode material. During the development of the MMC, SiCp was pre—heated at 500°C for up to 20 minutes. Aluminum alloy is then started to melt in a furnace at 700°C and mixed thoroughly for up to 20 minutes at 300 RPM. The pre—heated SiCp was then slowly added into the liquid aluminum. To achieve a uniform dispersion of SiCp in molten aluminum alloy, the molten circuitry was agitated at 300 RPM for up to 25 minutes. Afterwards, the melted matrix is injected into the mold to create the specified sized cast. The work-sample in consideration is rectangular in form (20X15X15mm). As tools, a multi-hole slotted cylinder was utilized to provide high-velocity air via the electrodes. Table 2.1 shows the chemical composition of the selected workpiece. Throughout the test, the material's hardness stays unchanged (50HRC). Figure 2.1b depicts the experimental setup employed for the current work.

2.2.2 TOOL DESIGN

Suitable tool design was chosen to provide continuous compressed air flow. The tool specifications of 9 mm diameter and 75 mm length were chosen to remove the most heat from the tool tip. The employment of the multi-hole slotted electrodes enabled the supply of compressed air via arc opening.

The multi-hole slotted electrodes offer various benefits over the tubular electrodes. This is due to the rotations of the multi-hole slotted electrodes improving flushing efficacy by providing a greater escape of gaseous dielectric and degraded fragments from the discharge domain, culminating in bettered machining productivity. Using conventional electrode throughout a pure drilling operation inhibits the production of gudgeons in the core of the borehole. On the tool's surface, 1.5-mm diameter holes were drilled at 4 mm, 5 mm, and 6 mm pitch circle

TABLE 2.1
The Chemical Constituents of the Work Sample

Constituent	Al	C	Si	Ni	O	Supplementary
Wt. (%)	83.42	9.33	4.26	0.89	1.02	1.08

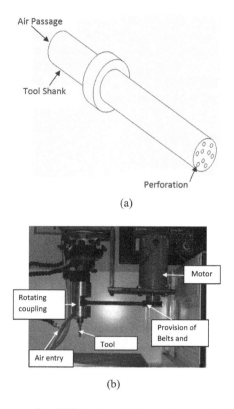

FIGURE 2.1 (a) Diagram of multi-hole slotted electrode and (b) pictorial view of experimental setup.

diameter. Regardless of how the machining was improved with this configuration, frequent the formation of electric arc and short-circuiting occurred due to poor ejection of debris from electrodes gap. As a result, three apertures of size 2 mm 2 mm were made throughout the tool's surface to address such issues. Figure 2.1a depicts the multi-hole slotted tool.

2.2.3 Experimental Procedure

An EDM machine was used for the air-assisted EDD with the multi-hole slotted tool. The machining duration was appropriately decided and set at 15 minutes for each trial. Throughout all trials, inexpensive EDM oil was employed as a dielectric. Five process factors were chosen for the exploratory work: pulse duration, discharge current, duty cycle, electrode speed, and discharge air pressure. Eventually, four factors were employed during the REDD, and all five factors were taken into account for AAEDD. Table 2.2 depicts the array of machining factors

TABLE 2.2
Factors and Their Levels in the Process

Parameters	Levels				
	−2	−1	0	1	2
Discharge current (I_p) (A)	3	4	5	6	7
Pulse on time (T_{on}) (μs)	100	200	300	400	500
Duty cycle (DC)	0.52	0.58	0.64	0.70	0.76
Tool rotation (rpm)	100	200	300	400	500
Air pressure (AP) (mm Hg)	3	5	7	9	11

used during the investigation. Throughout the studies, an open circuit voltage of 60V was kept constant.

The electrode wear ratio was calculated as wear down tool weight ratio to wear down workpiece weight after machining [23] and is given here:

$$\text{Electrode wear ratio}\left(\%\right) = \frac{\text{Eroded tool weight}}{\text{Eroded workpiece weight}} \times 100 \tag{1}$$

Material removal rate was found as fraction of weight of machined sample to overall machining time [23] as stated here:

$$\text{Material removal rate}\left(\text{mg/min}\right) = \frac{\text{Weight of eroded sample}}{\text{Machining time}} \tag{2}$$

The machined workpiece was cleansed in acetone. A weighing balance machine having least count of 0.1mg was utilized for the weight estimation. So, to guarantee a precise machining duration counts, electronic clock (exactness of 0.1 seconds) was employed for the present study.

2.3 DATA ANALYSIS ON THE EXPERIMENTAL OBSERVATIONS

In the present study, Central Composite Rotatable Design (CCRD) is utilized to plan the experimentation because it has the potential of predicting quadratic and interaction effects of different process factors on the process response. Tables 2.3 and 2.4 compare the computed values of each trial for MRR and EWR for the REDD and AAEDD processes, respectively. For the AAEDD, 32 experiments were carried out in accordance with the experimental design and 31 experiments were performed for REDD. The REDD process utilized four process parameters with five levels, whereas the AAEDD process utilized five parameters with five levels (the fifth variable is air pressure) for experiment design. Regression analysis of the data has been done to develop the models, which set up the relationship between input parameters

TABLE 2.3

The Estimated Magnitude of the Outputs Pertaining to the REDD Process Design of the Experiment

Exp. No.	Discharge Current (A)	Pulse on Time (T_{on})	Duty Cycle (DC)	Tool Speed (RPM)	MRR (mg/min)	EWR (mg/min)
1	6	200	0.58	400	11.13	1.89
2	7	300	0.64	300	14.91	2.23
3	6	400	0.70	400	9.07	1.93
4	5	300	0.64	300	8.19	1.85
5	4	400	0.58	400	6.16	1.70
6	5	300	0.52	300	6.97	1.59
7	4	200	0.58	200	7.39	1.56
8	6	400	0.58	200	9.59	1.75
9	4	400	0.58	200	5.30	1.83
10	4	200	0.58	400	6.60	1.86
11	3	300	0.64	300	4.04	1.47
12	5	300	0.64	300	7.94	1.82
13	5	100	0.64	300	14.22	1.89
14	5	300	0.64	300	8.91	1.83
15	6	200	0.70	400	14.28	2.17
16	4	200	0.70	400	8.58	1.72
17	5	300	0.64	100	8.01	1.69
18	6	400	0.58	400	8.08	1.87
19	6	400	0.70	200	10.21	1.90
20	6	200	0.70	200	14.57	2.14
21	4	400	0.70	400	4.29	1.82
22	5	500	0.64	300	5.08	1.71
23	5	300	0.64	300	8.67	1.81
24	4	400	0.70	200	5.19	1.68
25	5	300	0.76	300	8.87	1.89
26	5	300	0.64	500	5.92	1.86
27	5	300	0.64	300	8.14	1.79
28	5	300	0.64	300	8.19	1.78
29	6	200	0.58	200	12.20	2.02
30	4	200	0.70	200	8.67	2.09
31	5	300	0.64	300	7.18	1.82

and process responses. After removing all inconsequential factors, the required regression models based on the trial results of the AAEDD process are given by the following equations.

$$MRR = -43.6 + (1.31 \times I) - (0.0219 \times T_{on}) + (74.8 \times DC) + (0.0155 \times rpm)$$
$$+ (4.01 \times AP) - (0.000031 \times rpm^2) - (0.0495 \times AP^2)$$
$$+ (0.221 \times I \times AP) - (5.48 \times DC \times AP) \tag{4}$$

TABLE 2.4
The Estimated Magnitude of the Outputs Pertaining to the AAEDD Process Design of the Experiment

Exp. No.	Discharge Current (A)	Pulse on Time (T_{on})	Duty Cycle (DC)	Tool Speed (RPM)	Air Pressure (AP) (mm-Hg)	MRR (mg/min)	EWR (mg/min)
1	6	200	0.58	400	9	19.31	1.80
2	7	300	0.64	300	7	19.50	2.04
3	6	400	0.70	400	9	14.16	1.74
4	5	300	0.64	300	7	12.76	1.64
5	4	400	0.58	400	9	6.57	1.54
6	5	300	0.52	300	7	8.29	1.50
7	4	200	0.58	200	9	8.97	1.37
8	6	400	0.58	200	9	14.84	1.54
9	4	400	0.58	200	5	4.66	1.43
10	4	200	0.58	400	5	6.92	1.62
11	3	300	0.64	300	7	5.45	1.38
12	5	300	0.64	300	7	12.75	1.64
13	5	100	0.64	300	7	17.57	1.75
14	5	300	0.64	300	7	11.84	1.65
15	5	300	0.64	300	11	12.35	1.61
16	6	200	0.70	400	5	18.91	2.42
17	4	200	0.70	400	9	9.79	1.37
18	5	300	0.64	100	7	9.03	1.55
19	6	400	0.58	400	5	7.08	1.73
20	6	400	0.70	200	5	11.44	1.75
21	6	200	0.70	200	9	17.84	2.02
22	4	400	0.70	400	5	7.28	1.78
23	5	500	0.64	300	7	7.39	1.58
24	5	300	0.64	300	3	7.29	1.81
25	4	400	0.70	200	9	9.70	1.66
26	5	300	0.76	300	7	14.80	1.84
27	5	300	0.64	500	7	9.77	1.78
28	5	300	0.64	300	7	11.69	1.64
29	5	300	0.64	300	7	11.63	1.68
30	6	200	0.58	200	5	13.19	1.91
31	4	200	0.70	200	5	11.93	1.49
32	5	300	0.64	300	7	11.45	1.65

$$EWR = -1.27 - (0.039 \times I) + (0.00450 \times T_{on}) - (1.79 \times DC) + (0.000270 \times rpm)$$
$$- (0.0925 \times AP) + (0.0164 \times I \times I) + (0.00169 \times AP \times AP) \qquad (5)$$
$$- (0.00119 \times I \times T_{on}) + (0.639 \times I \times DC) + (0.000139 \times T_{on} \times AP)$$

Analysis of variance (ANOVA) has been applied to explore competency of the developed model. The predictive model's F-ratio was evaluated in comparison to the

TABLE 2.5
ANOVA Table for MRR

Source	DF	Seq. SS	MS	F	P	R²	
Regression	9	518.892	25.094	42.29	0	0.987	$F^{standard}_{(0.05,9,22):} = 2.90$
Linear	5	452.203					
Square	2	37.202					$F^{regression} > F^{standard}_{(0.05,9,22)}$
Interaction	3	27.563					
Residual error	22	25.709	0.879				$F^{standaed}_{(0.05,17,22):} = 2.96$
Lack-of-Fit	17	27.978		2.48	0.059		$F^{lack-of-fit} < F^{standard}_{(0.05,17,22)}$
Pure error	5	1.679					The model is reliable.
Total	31	525.641					There seems to be a trivial lack of fit.

TABLE 2.6
ANOVA Table for EWR

Source	DF	Seq. SS	MS	F	P	R²	
Regression	9	1.42532	0.07103	36.29	0	0.981	$F^{standard}_{(0.05,9,22):} = 2.762$
Linear	5	1.03175					
Square	1	0.17519					$F^{regression} > F^{standard}_{(0.05,9,22)}$
Interaction	3	1.40057					
Residual error	22	0.51626	0.042				$F^{standaed}_{(0.05,9,22):} = 2.762$
Lack-of-Fit	17	0.47093		1.47	0.110		$F^{lack-of-fit} < F^{standard}_{(0.05,17,22)}$
Pure error	5	0.04533					The model is reliable. There seems to be a trivial lack of fit.
Total	31	4.74247					

tabulated F-value for each confidence interval. Tables 2.5 and 2.6 show the ANOVA of the second-order model of the AAEDD. The statistical equations of MRR and EWR are represented by equations 3 and 4. The assessment of "Prob>F" for the proposed model is less than 0.05. (95 percent confidence). Henceforth, it uncovers that the process factors in the model have a significant effect on the process output.

Figures 2.2 show the impact of the pulse duration on MRR in AAEDD. It was caused by a decrease in MRR as the pulse duration increased. It was because of the development of spark column with respect to the time during which energy was supplied prompting the decrease in energy density that brought about a decrease of MRR with an expansion in pulse on time. This could be since the presence of air in AAEDD inflicted exothermic reactions and triggered extra heat besides the heat of plasma, resulting in an improved MRR in AAEDD. Furthermore, MRR is decreased when the delivery of air and tool rotation gets too high. It was most likely due to increased flow of air and tool rotation, which stirred up the plasma passage. Figure 2.3 depicts the contributor of each factor in the MRR models. Figure 2.3 illustrates that the discharge current, duty cycle, and tool rpm are all essential aspects

Electrical Discharge Drilling of Al-SiC

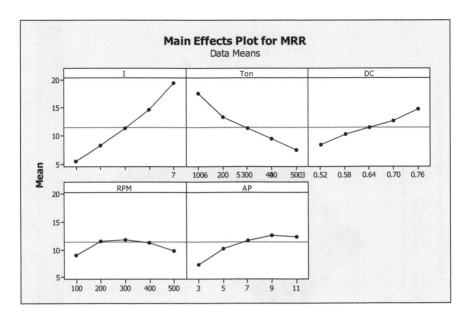

FIGURE 2.2 Main effect plots for MRR.

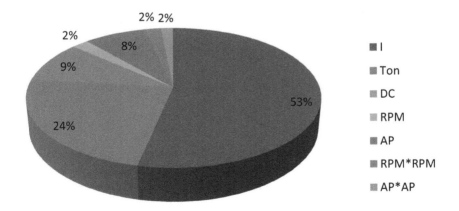

FIGURE 2.3 Contribution (%) of process factors towards MRR.

influencing the EWR. Besides the pulse duration and duty cycle, the discharge current is perceived to be the important factor influencing the MRR.

2.3.1 Response Surface of MRR

Figure 2.4 depicts the MRR for discharge current and discharge air pressure. It is possible that a high discharge current led to increased MRR. This is because the

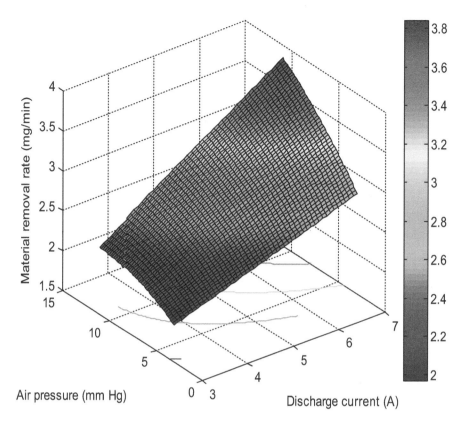

FIGURE 2.4 AAEDD interaction plot of discharge current and discharge air pressure on MRR.

discharge energy in the machining zone has increased, resulting in further evaporation and melting of the work piece. From the graph, it could be observed that MRR improved with an increment in discharge air pressure reaching its maximum value and then started decreasing at a low level of discharge current. However, at a high level of the discharge current, a continuous increase in MRR with air pressure was obtained. This can be accounted for the reason that the flushing efficacy of the procedure improved with rise in discharge air pressure, which contributed to high MRR [12].

Figure 2.5 depicts the MRR for discharge air pressure and duty cycle. MRR increases with increasing discharge air pressure and duty cycle, as shown in the graph. This was very likely since the greater duty cycle value encouraged high spark energy, which increases evaporation and extraction of the work material. The increased air pressure aided flushing of disintegrated material from the discharge gap, leads to a higher MRR.

Electrical Discharge Drilling of Al-SiC

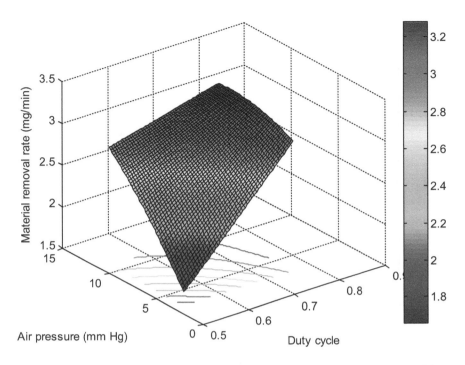

FIGURE 2.5 The effect of duty cycle and discharge air pressure on MRR is depicted by an interaction plot.

2.3.2 Response Surface of EWR

The impression of a discharge current on EWR for the AAEDD process is manifested in Figure 2.6. It is clear that the increase in discharge current rapidly increased the temperature of the plasma. It increased the overheating and vaporization of the tool, resulting in a higher in EWR. Figure 2.6 depicts how the EWR reduced as air pressure increased in the AAEDD. The passage of air via the tool holes brought away the heat, lowering the temperature of the tool tip. The effect of cooling further enhanced through pressure rise. As a result of the increased air pressure in the AAEDD process, the EWR was reduced. Figure 2.7 depicts the proportion of each factor in the EWR models. According to the pattern, the discharge current, pulse duration, and duty cycle are all important aspects that affect the EWR. Apart from pulse duration and duty cycle, the discharge current is noticed to be a more critical aspect influencing the EWR.

The interaction effect among the discharge current and pulse-on time on EWR is depicted in Figure 2.8. From the plot, it could be observed that the combination of high current and low pulse duration lead to an excessive EWR. A greater current led to increment in discharge energy and, as a result, a high temperature. Because of this intense heat, there occurs an overheating of the electrode which increased the EWR.

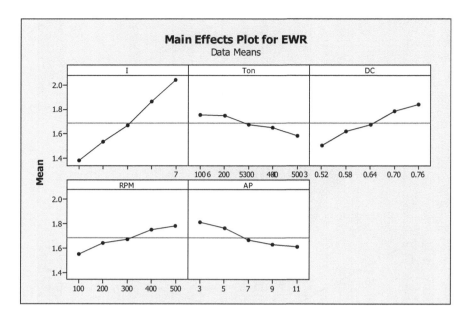

FIGURE 2.6 EWR main effect plot lines.

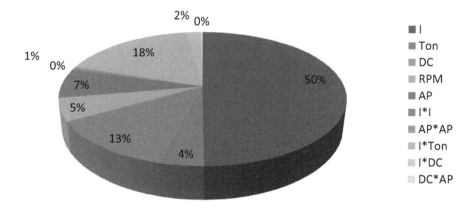

FIGURE 2.7 Role in contributing of input parameters to EWR.

The dark carbon produced by the deformation of mineral oil that adheres to the working electrode at longer pulse durations. The development of this kind of layer increases the electrode's resistance to wear, leading to reduced EWR [10].

The EWR response surface for a pulse duration and release air pressure is shown in Figure 2.9. The graph clearly demonstrates that combining longer pulse duration with a greater downstream air pressure results in a significant drop in EWR. Increased release air pressure facilitated cooling of the tool. As the pulse duration is expanded, the temperature of the electrode tends to decrease caused by the spread of the plasma channel.

Electrical Discharge Drilling of Al-SiC

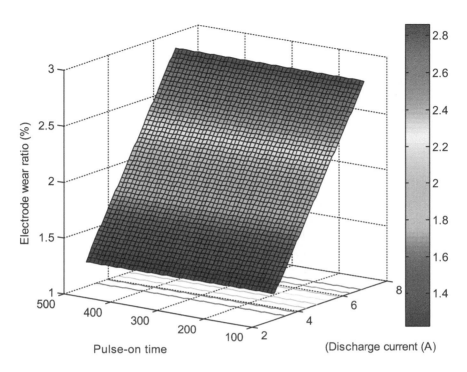

FIGURE 2.8 Interface graph between discharge current and pulse duration on EWR.

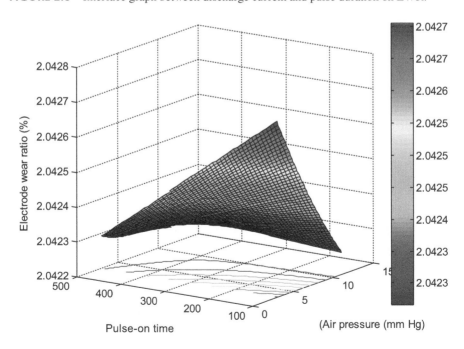

FIGURE 2.9 Interface graph between discharge air pressure and pulse duration on EWR.

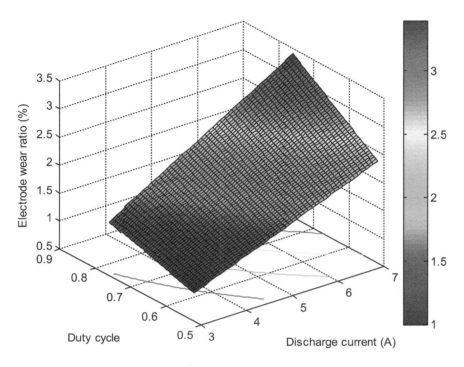

FIGURE 2.10 Interface graph between discharge current and duty cycle on EWR.

Figure 2.10 depicts the EWR response surface for duty cycle and discharge current. It might be deduced that the high EWR was caused as a result of a large duty cycle and a higher flow current. Considerable discharge energy and impetuous force were characterized by greater duty cycle and discharge current, resulting in larger sized carters on the workpiece and a rise in EWR with a rise in duty cycle and discharge current [22]. From the plot, it was also observed that the EWR enhanced with the increase in discharge current at all levels of duty cycle.

2.4 RESULTS AND DISCUSSION

2.4.1 The Influence of Control Factors on MRR

Figure 2.11 (a) depicts the influence of discharge current during the AAEDD and REDD processes MRR increased as discharge current increased. This was most likely caused by an increment in discharge energy in the machining region that further enhanced material decomposition and evaporation. According to the graph (refer to Figure 2.8a), MRR for AAEDD was comparatively better than MRR obtained through the REDD. This is because the energy content of the spark increased as the discharge current increased, leading to a huge crater quantity. The perforation on the other hand, aided in the efficacious propagating of dielectric, as it accelerated cleanup efforts from the electrodes gap. Furthermore, the oxygen level of the air

Electrical Discharge Drilling of Al-SiC

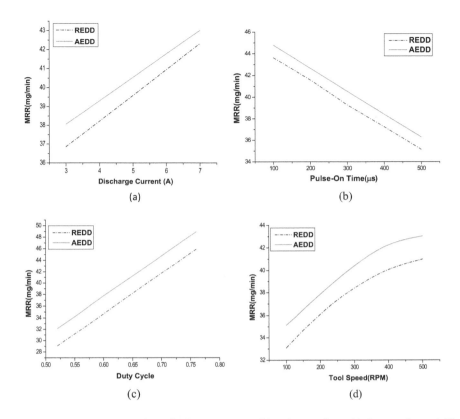

FIGURE 2.11 Influence of (a) discharge current, (b) pulse-on time, (c) duty cycle, and (d) tool speed on MRR in REDD and AAEDD.

triggers an exothermic reaction that provides further heat for specimen melting in the AAEDD procedure [23].

The effect of pulse-on time, during AAEDD and REDD processes on the material removal rate has been shown in Figure 2.11(b). The material removal rate was found to decrease as the pulse duration increased. This was most probably owing to the assertion that at high pulse duration, large decayed particulates were produced, and these pieces accrued inside the machining zone and approached the tool surface, resulting in short-circuiting. As a consequence of the greater non-cutting machining period, though at a high pulse-on time, a reduced MRR is noted [24]. A lengthy pulse time during REDD resulted in plasma channel expansion, resulting in a lower energy density. As a result of significantly lower melting and vaporization of the specimen material, reduced MRR at extended pulse duration is observed [25]. Furthermore, the figure shows that AAEDD has a higher MRR as compared to REDD. This was most likely due to the increased expulsion of evaporated and liquefied substance from the discharge gap caused by the stream of pressurized air via a proposed tool.

The effect of tool rotation on the MRR during REDD and AAEDD has been shown in Figure 2.11(c). The pattern reveals that in both processes, at greater tool rotation values, the MRR reduced. It's most likely owing to enhanced turbulence caused by high rotation speed, which stirred up the plasma channel. Increased tool speed resulted in significant centrifugal force, resulting in a vortex of dielectric fluid across the electrodes passage [26].

The effect of duty cycle on MRR during both processes has been shown in Figure 2.11(d). MRR was found to increase with a rise in duty cycle. It was the cause of the high spark energy freed at a higher duty cycle, which resulted in additional melting of workpiece material. The plot clearly shows that MRR obtained through AAEDD was larger than the REDD. It's most probably because the deployment of a multi-hole slotted tool that aided in the extraction of debris and heat from the spark region. As a consequence, it enhanced the flushing function, resulting in a high MRR.

2.4.2 EFFECT OF PROCESS PARAMETERS ON EWR

The tool electrode governs the cavity to be created in the EDM operation. A good EDM tool electrode ought to be highly conductive, have a reduced tool wear, and produce a higher surface finish. Figure 2.12(a) depicts the influence of discharge current on EWR for proposed processes. EWR increased with increasing discharge current in those procedures. This was most likely due to the fact that higher current produced higher discharge energy. As a result of the increased temperature in the machining zone, wear of the tool as well as MRR increased. As a result, the electrode wear ratio rose [25]. Furthermore, at large discharge current, a significant portion of eroded particulate amasses in the machining gap, increasing electrode wear [26]. The figure shows that EWR was lower in the AAEDD than in the REDD. This one was attributed to the rationale that as pressurized air was provided via the multi-hole slotted tool, the temperature of the tool surface decreased, and as a result, the crater formed during the AAEDD decreased on the tool.

Figure 2.12b presents the effect of pulse on time on EWR. It was discovered that EWR has an asymmetric relationship with a pulse on time. It could be attributed to the essence that as pulse duration increased, the size of the plasma channel widened, lowering the high temperature of the electrode exterior [24]. Furthermore, for extended pulse durations, the black carbon produced by petroleum oil crack propagation adhered to the electrode's exterior. The carbon layers deposited on the tool safeguard it from excessive tool wear [16]. Furthermore, the figure shows that EWR was smaller in AAEDD when weigh up to the REDD. It was most likely due to the fact that compressed air significantly lowered the temperature of the copper electrode, minimizing melting and vaporization of the tool and thus minimizing its wear.

The effect of duty cycle on EWR for AAEDD and REDD processes is depicted in Figure 2.12c. It demonstrated that EWR enhanced in undeviating proportion to duty cycle. Strong spark energy was emancipated as a result of a duty cycle increase, which inevitably resulted in a high EWR [26]. The graph indicated that the EWR in AAEDD has been lower than in the REDD. This has been due to the fact that the multi-hole slotted rotary tool utilized during the AAEDD improved flushing efficacy in the sparking area by simply eliminating debris and heat.

Electrical Discharge Drilling of Al-SiC

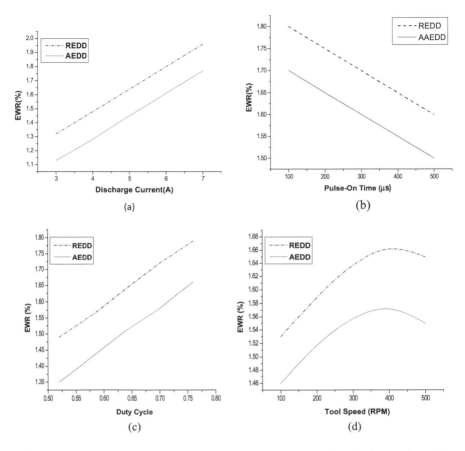

FIGURE 2.12 Influence of (a) discharge current, (b) pulse-on time, (c) duty cycle, and (d) tool rotation on EWR in REDD and AAEDD.

Figure 2.12d depicts the impact of tool speed on EWR for AAEDD and REDD. The pattern reveals that as tool rpm increased EWR gone up for both procedures. The magnitude of centrifugal force raised as the rotating speed of the electrode enhanced. The black carbon produced by dielectric crack propagation adheres to the electrode's exterior. This helps improve the tool's resistance to wear. However, due to increased centrifugal force, the stack of carbon drifted away from the exterior of the electrode, resulting in increased wear rate [27]. Besides that, increasing the rotary rpm of the tool is helpful to enhance the flushing efficacy of the discharge disparity as a result the portion of particulate in the machining zone was lowered, resulting in higher EWR [15]. Furthermore, the plot shows that EWR was higher in conventional REDD especially in comparison to AAEDD. It was most likely because compressed gas caused faster solidification of suspended carbon (on tool surface) due to dielectric breakdown. Moreover, the cooling of the electrode surface felicitated the formation of carbon layers. Hence, when compressed air was passed through tool the formation of carbon deposition increased.

2.4.3 Surface Morphology

Owing to the quick and frequent strike of elevated electron beam, the machined surface reaches elevated temperatures during EDM process. Due to the extreme elevated temperatures, the material quickly disintegrates, vaporizes, and cavities form on the electrode surface. When the severity of thermal stresses surpasses the fracture strength of the work material in an EDM procedure, crack initiation begins [20]. The SEM images as shown in Figure 2.13 illustrate the surface morphology of machined workpiece.

(i) Under, discharge current 3A, pulse on time of 300µs, duty cycle of 0.64, tool rotation 300rpm

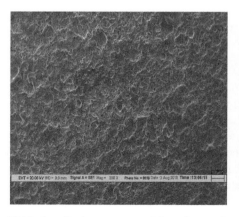 

(ii) Under, discharge current 7A, pulse on time of 300µs, duty cycle of 0.64, tool rotation 300rpm

FIGURE 2.13 Surface attributes of a machined component (A) AAEDD process (B) REDD process.

Figure 2.13 shows that parts machined using the REDD process have more micro pores, superficial ravines, and recast strands than work—piece machined using the AAEDD process. It can be seen (refer to Figures 2.13c and 13d) that the specimen machined by AAEDD has fewer fractures than the test specimen machined all through the REDD process. This is due to the fact that compressed air reduces temperature variation. Furthermore, SEM images show that AAEDD-machined sample have less recast layer than REDD-machined work sample. This is because the compressed air provides a better flushing, which removed the debris from the surface of the machined workpiece [24].

2.4.4 PROCESS OPTIMIZATION AND ACCURACY OF MODELS

The various calculated responses and parameters are often uncertain because of the experimental errors. Although the accuracy of the process responses was determined through estimation of the confidence interval, determined using relation $Y \pm \Delta Y$, where, ΔY is represented by

$$\Delta Y = t_{\alpha/2}, DF \sqrt{V_e} \qquad (6)$$

Here the process responses, EWR, MRR, and SR are denoted by Y. The level of the confidence interval is denoted by α and its value is taken as 0.05. The variance of error of the foreseen procedure output is presented by V_e.

The MRR and EWR ranges were approximated, and experiments were carried out to confirm the ranges at different parameter combinations. The experimental values were discovered to be lower than the regression model range evaluation (see Table 2.7).

2.4.5 GENETIC ALGORITHM-BASED MULTI-OBJECTIVE OPTIMIZATION

The algorithm is proposed multi-objective optimization technique, that is accessible in MATLAB® 12, had been utilized to enable optimal MRR and lower EWR. It aids

TABLE 2.7
Confirmatory Experiments

S. No.	Process factors					MRR (mg/min)		EWR (mg/min)	
	I_p	Ton	DC	RPM	AP	Experiment	Anticipated	Experiment	Anticipated
1	6	300	0.76	400	7	11.40	11.04 ± 1.66	1.35	1.40 ± 0.25
2	5	100	0.70	200	5	12.03	10.73 ± 1.66	1.65	1.83 ± 0.24
3	4	300	0.52	500	11	16.32	15.04 ± 1.66	1.72	1.86 ± 0.24

TABLE 2.8

Optimal Factor Values for MRR and EWR

I_p	Ton	DC	(RPM)	AP		MRR (mg/min)		EWR (mg/min)	
					Anticipated	Experiment	Anticipated	Experiment
5.5	244	0.53	260	18.5	24.01 ± 1.88	25.30	1.40 ± 0.24	1.35

in the resolution of bound limitations concerns. RSM regression equations are used as activation function in MATLAB to generate the objective functions, and RSM design variations were employed as acceptable limits for five process variables. As a criterion, the following established statistical description symbolized by formula (5), has been utilized.

$$\text{Minimize} = f\ (EWR, (-MMR)) \tag{5}$$
$$\text{Subjected to } 3 \leq X1 \leq 7;\ 100 \leq X2 \leq 500;\ 0.52 \leq X3 \leq 0.76;\ 100 \leq X4 \leq 500;\ 3 \leq X5 \leq 11$$

Table 2.8 shows the optimal variable values that contribute to better MRR and lower EWR. The model was validated by running three distinctive types of experimentation at an optimum level of factors. Table 2.8 shows the average value of the experimental outcome from these three trials under optimal environments. The difference between anticipated and experiment values is within justifiable bounds.

Table 2.8 demonstrates that the optimal value of MRR and EWR were in the sequence of 25.30 and 1.35 during the AAEDD process of Al-SiC composite. These optimal parametric values are extremely helpful during the EDM process in order to improve process performance and improving tool topology viability for a prolonged period of time.

2.5 CONCLUSIONS

In this work, the effects of a hybrid process as well as impacts of process parameters on output responses including MRR and EWR during EDD of Al-SiC ceramic composite was investigated. An air-assisted rotary tool electrode was effectively utilized to machine Al-SiC specimens in the current work. Furthermore, this study investigated the effectiveness of an air-assisted multi-hole slotted tool to improve the flushing performance of the machining region, thereby enhancing the efficiency of removal of material during EDM drilling.

- A comparison of response characteristics for a solid rotary tool and an air-assisted multi-hole slotted rotary tool is addressed. In comparison to the REDD, the AAEDD process gained higher MRR and reduced EWR.
- Statistical models were also developed for predicting the responses like EWR, and MRR during AAEDD process. Experimental findings suggested that discharge current, pulse on time and duty cycle, notably affected the

MRR, whereas discharge current, pulse on time and tool rotation significantly affected the EWR.

- The Surface analysis showed that recast layers were more pronounced on specimen machined by AAEDD in comparison to REDD process. Moreover, in this respect, it has been noted that surface cracks induced during the AAEDD process were less relative to the REDD process.
- The ideal machining conditions in AAEDD (discharge current 5.5A, pulse on time 244µs, duty cycle 0.53, tool speed 260 rpm, and gas pressure 18.5 mm of Hg) produced MRR and EWR comparable to 25.30 mg/min and 1.35, respectively.

REFERENCES

[1] K.H. Ho and S.T. Newman, State of the art electrical discharge machining (EDM). *Int. J. Mach. Tools Manuf.* 43 (2003), 1287–1300.

[2] N.K. Singh, P.M. Pandey, K.K. Singh, and M.K. Sharma, Steps towards green manufacturing through EDM process: A review. *Cogent Eng.* 3 (2016), 1272662.

[3] C.C. Wang and B.H. Yan, Feasibility study of rotary electrical discharge machining with ball burnishing for $Al_2O_3/6061Al$ composite. *Int. J. Mach. Tools Manuf.* 40 (2000), 1403–1421.

[4] B. Mohan, A. Rajadurai, and K.G. Satyanarayana, Effect of SiC and rotation of electrode on electric discharge machining of Al-SiC composite. *J. Mater. Process. Tech.* 124 (2002), 297–3042.

[5] P. Kuppan, A. Rajadurai, and S. Narayanan, Influence of EDM process parameters in deep hole drilling of Inconel 718. *Int. J. Adv. Manuf. Tech.* 38 (2008), 74–84.

[6] H. Hocheng, W.T. Lei, and H.S. Hsu, Preliminary study of material removal in electrical discharge machining of SiC/Al. *J. Mater. Process. Technol.* 63 (1997), 813–818.

[7] Y.W. Seo, D. Kim, and M. Ramulu, Electrical discharge machining of functionally graded 15–35 Vol% SiCp/Al composites. *Mater. Manuf. Process.* 21 (2006), 479–487.

[8] K.M. Patel, P.M. Pandey, and V.P. Rao, Study on machinability of Al2O3 ceramic composite in EDM using response surface methodology. *J. Eng. Mater. Tech.* 133 (2011), 21004–21009.

[9] S. Suresh Kumar, M. Uthay Kumar, S.T. Kumaran, and P. Parameswaran, Electrical discharge machining of Al(6351)—SiC—B_4C hybrid composite. *Mater. Manuf. Process.* 29 (2014), 1395–1400.

[10] R. Karthikeyan, L. Narayanan, and R.S. Naagarazan, Mathematical modeling for electric discharge machining of Al-SiC particulate composites. *J. Mater. Process. Tech.* 87 (1999), 59–63.

[11] I. Puertas and C.J. Luis, A study of optimization of machining parameters for electrical discharge machining of boron carbide. *Mater. Manuf. Process.* 19 (6) (2004), 1041–1070.

[12] I. Puertas and C.J. Luis, Modeling the manufacturing parameters in electrical discharge machining of siliconised silicon carbide. *Proc. Instn. Mech. Engineers Part B: J. Eng. Manuf.* 217 (2003), 791–803.

[13] S. Dhar, R. Purohit, N. Saini, and A. Sharma, Mathematical modeling of electric discharge machining of cast Al-4Cu-6Si alloy-10 wt. % SiCp composites. *J. Mater. Process. Technol.* 193 (2007), 24–29.

[14] E. Aliakabari and H. Baseri, Optimization of machining parameters in rotary EDM process by using the Taguchi method. *Int. J. Adv. Manuf. Tech.* 62 (9) (2012), 1041–1053.

[15] L. Gu, L. Li, and W. Zhao, Electrical discharge machining of Ti6Al4V with a bundled electrode. *Int. J. Mach. Tools Manuf.* 53 (2012), 100–106.

[16] A. Singh, P. Kumar, and I. Singh, Electrical discharge drilling of metal matrix composites with different tool geometries. *J. Process Mech. Engg.* (2013). doi: 10.1177/0954405413484726.

[17] N.K. Singh, P.M. Pandey, and K.K. Singh, EDM with air assisted multi-hole rotating tool. *Mater. Manufac. Processes* 31 (14) (2006), 1872–1878.

[18] M. Yoshida, Y. Ishii, and T. Ueda, Study on electrical discharge machining for cemented carbide with non-flammable dielectric liquid: Influence of form of oxygen supplied to dielectric liquid on machining. *Proc. IMechE Part B: J. Engg. Manufac.* (2017). doi: 10.1177/0954405417706995.

[19] L. Selvarajan, C.S. Naryanan, R. Jeyapaul, and M. Manohar, Optimization of EDM process parameters in machining Si_3N_4—TiN conductive ceramic composites to improve form and orientation tolerances. *Meas.* 92 (2016), 114–129.

[20] A. Torres, C.J. Luis, and I. Puertas, EDM machinability and surface roughness analysis of TiB_2 using copper electrodes. *J. Alloys Compd.* 690 (2017), 337–347.

[21] M. Tanjilul, A. Ahmed, A.S. Kumar, and M. Rahman, A study on EDM debris particle size and flushing mechanism for efficient debris removal in EDM-drilling of Inconel 718. *JMPT* 255 (2018), 263–274.

[22] V. Srivastava and P.M. Pandey, Effect of process parameters on the performances of EDM process with ultrasonic assisted cryogenically cooled electrode. *J. Manuf. Process.* 14 (2012), 393–402.

[23] N.K. Singh, P.M. Pandey, and K.K. Singh, Experimental investigations into the performance of EDM using argon gas-assisted perforated electrodes. *Mater. Manuf. Proc.* (2016). doi: 10.1080/10426914.2016.1221079.

[24] H. Beravala and P.M. Pandey, Experimental investigations to evaluate the effect of magnetic field on the performance of air and argon gas assisted EDM processes. *J. Manuf. Process.* 34 (2018), 356–373.

[25] N.K. Singh, P.M. Pandey, and K.K. Singh, A semi-empirical model to predict material removal rate during air-assisted electrical discharge machining. *J. Braz. Soc. Mech. Sci. Eng.* 41 (2019), 122.

[26] N.K. Singh, Y. Singh, and A. Sharma, Experimental investigation and statistical modeling of air-assisted EDM of Al–SiC composite with an improvised tool. *J. Braz. Soc. Mech. Sci. Eng.* 42 (2020), 312.

[27] K.D. Chattopadhyaya, S. Verma, and P.C. Satsangi, Development of empirical model for different process parameters during rotary electrical discharge machining of copper–steel (EN-8) system. *J. Mater. Process. Tech.* (2009), 1454–1465.

3 Web Buckling Investigation of Direct Metal Laser Sintering-Based Connecting Rod with Hexagonal Perforations

Gulam Mohammed Sayeed Ahmed,
Mengistu Gelaw Perumall, Janaki Ramulu,
Belay Brehane, Devendra Kumar Sinha,
and Satyam Shivam Gautam

3.1 Introduction 52
3.2 Hexagonal Perforations...53
    3.2.1 Constructal Hypothesis...53
3.3 Regression Analysis..55
3.4 Materials and Methods 56
    3.4.1 Three-Dimensional Printing Metals.....................................56
        3.4.1.1 Aluminum Al-Si-10Mg 56
        3.4.1.2 Stainless Steel 17-4PH 56
        3.4.1.3 Stainless Steel 316L 57
        3.4.1.4 Titanium Ti64 57
    3.4.2 Direct Metal Laser Sintering (D-M-L-S) Technology 58
3.5 CAD Geometry with Hexagonal Perforations in Connecting Rods..............60
    3.5.1 Criteria in the Lightweight Web Design with Hexagonal Perforations..60
    3.5.2 Buckling Criterion 60
3.6 Result and Discussions 63
    3.6.1 Structural Analysis 63
    3.6.2 Modal Analysis and Frequency Response............................63
    3.6.3 Buckling Analysis...64
    3.6.4 Variation of Normalized Limiting Stress with Geometric Characteristics 66

DOI: 10.1201/9781003220237-3

3.6.4.1	AISiMg0.6	66
3.6.4.2	Stainless Steel 17_4PH	67
3.6.4.3	Stainless Steel 316L	71
3.6.4.4	Titanium Ti64	71
3.7	Conclusions	73
3.8	Acknowledgement	75
3.9	Conflicts of Interest	75
References		75

3.1 INTRODUCTION

An important component of an automobile engine is a connecting rod that has mass volume production. Depending on the requirement of an internal combustion engine each connecting rod is associated with the number of engine cylinders furthermore plays a critical link between the crankshaft and piston. Connecting rods are mostly used in single cylinder automotive engines of modern bikes and scooters. They are a versatile component that transmits the cyclic reciprocal motion of the engine piston along with revolution of the crank-shaft by means of a connecting rod [1–2]. Mohammad-Reza et al. [3–4] revealed the effects of dynamic combustion load on an engine's connecting rod and its optimum design parameters. They also determined the sensitivity of buckling in reduced weight of connecting rods. The stress-induced sensitivity in buckling behavior is comparatively more than yield and fatigue stresses. The other studies reported the connecting rod with design modifications and parametric study optimization by applying a finite-element (FE)-based procedure [5–6]. The weight reduction criteria with aluminum as rod material over forged steel discussed and the FE based approach was applied in structural deformation data analysis to avoid manufacturing related barriers and issues associated with the complex geometrical and boundary-mesh conditions. Hippliti R and Lu PC et al [7–9] analyzed the stress data from FE based procedures in the design optimization along with the fatigue life calculation for assessment of cyclic loading during engine running conditions. Other parametric optimization studies based on fatigue stress data has been reported [10–14]. In the present research based on the previous literature surveys on material selection for fabrication of connecting rods, an attempt has been made to apply the mechanical properties of D-M-L-S-based 3D metal printing in simulation by utilizing the FE-method-based software SIMSOLID. Most recent studies explore the abundant capabilities of additive manufacturing (Add-Mfg.) in product development by different 3D printing metals with feasibilities in mass production [14–16]. S. Ford et al. and L. E. Murr et al. [17–20] presented the advantages and challenges in sustainability of Add-Mfg. G.M.S. Ahmed et al. [21] reported design and discussed FE thermal analysis of a direct metal laser sintering of maraging steel. In other research work development of Light weight structures and application of cellular materials as load bearing mechanical components developed by 3D printing (3DP) has been presented [22–24]. The web of the connecting rod is a critical section of the component that experiences cyclic fatigue loading and experience buckling effect due to heavy gas pressure acting at the big end of connecting rod. In this present work FE based Simulation of buckling of the connecting rod has been studied by incorporating hexagonal perforations (HP) on the web and the buckling stress results are compared with and without HP. Evaluation of buckling stress in connecting rods attracts the manufacturers of automotive components. Some studies explained the buckling effect with substantial weight and area reduction of web by incorporating circular and hexagonal perforation in castellated beams [25–31]. The investigation of buckling sensitivity analysis of connecting rods performed on ultra-fine grained material AA2618. The numerical analysis was carried out through the finite element method (FEM) tool ANSYS [32]. The investigation on buckling analysis and stability of compressed low carbon steel rods was carried out and the critical stresses were determined [33].

In the present research an attempt has been made to fill the research gap of reduction in web buckling of connecting rods by reducing the mass of the connecting rod by considering D-M-L-S-based metals. By reducing the overall mass of connecting rod leads to the reduction of rotational and translational mass. Furthermore there are three merits in the engine performance: (1) As load transmitted from connecting rod decreases leads to increase in durability of crank shaft, (2) Enhancement in fuel efficiency due to reduction in mass of connecting rod, and (3) based upon the structural deformation analysis probability of failure will be less with minimum design volume.

3.2 HEXAGONAL PERFORATIONS

3.2.1 CONSTRUCTAL HYPOTHESIS

Slenderness is an important geometrical characteristic of tapered I-section webs. When a web of connecting rod subjected to cyclic loading during an axial compression critical mechanical behavior occurs called buckling leads to displacements out-of-domain. In the present investigation a buckling behavior in the web of connecting rods with hexagonal perforations were studied. Transverse hexagonal perforations have been incorporated on the web and FE-based numerical simulation using the SIMSOLID software tool to evaluate the critical buckling-load and normal limiting stress (NLS) with equal and increasing HPs on webs of connecting rods. The constructal hypothesis was applied to evaluate the effects of the dimensions of hp/lp where hp and lp are width and length, respectively, on buckling phenomena. The volume fraction (ψ) of perforations is a versatile parameter to evaluate buckling stress with and without perforations, where volume fraction defined as ratio of volume of the connecting rod with perforations to the volume of the solid connecting rod. The variation in NLS and critical buckling stress with respect to hp/lp was studied by considering different dimensions of hexagonsas given in Table 3.1. The design methodology includes the constructal hypothesis, which states that a finite dimensional domain survives with respect to time and its geometric configuration facilitates the boundary conditions to be accessed through the domain. The domain possesses (1) basic properties such as the boundary conditions type that influence the domain with content like mass and heat transfer, laminar or turbulent flow, force or Stress,

TABLE 3.1
Dimensions Selected for the Constructal Hypothesis

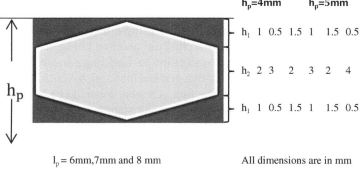

	$h_p=4$mm			$h_p=5$mm		
h_1	1	0.5	1.5	1	1.5	0.5
h_2	2	3	2	3	2	4
h_1	1	0.5	1.5	1	1.5	0.5

$l_p = $ 6mm, 7mm and 8 mm   All dimensions are in mm

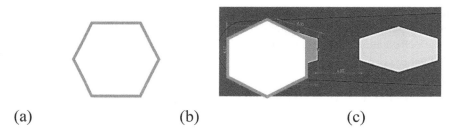

(a)   (b)   (c)

FIGURE 3.1 (a) Longitudinal HP, (b) transverse HP, and (c) dimensions of HP on the web of the connecting rod.

acoustic phenomenon etc., and (2) the geometrical features or design that describes the domain shape. The constructal hypothesis can be utilized in complex engineering problems that include objective parameters, boundary conditions, degrees of freedom and constraints to determine the influence of these factors on domain analysis. The dimensions of the perforations with several equivalent geometric data sets suggested by the constructal hypothesis in web design application were compared with solid, equal HP and increasing HP. The degree of freedom considered is h_p/l_p and the web of the connecting rod is considered as tapered plate of an I-section beam with h_p and l_p dimensions, given in Table 3.1. Transverse HP has been selected as the force boundary condition to act on the large side of the tapered web and provides more rigidity on the edges of the HPs as shown in Figure 3.1. Longitudinal HP promotes stress concentration factors at corner edge points apposing force direction, i.e., at the larger side of the web of the connecting rod. Another disadvantage is buckling sensitivity at the sharp corner edge point on right side of the perforation. The dimensions of the perforations with several equivalent geometric data sets suggested

Web Buckling Investigation

by the constructal hypothesis in web design applications were compared with solid, equal, and increasing HP.

The degree of freedom considered is h_p / l_p and the web of the connecting rod is considered as the tapered plate of an I-section beam with h_p and l_p dimensions, as given in Table 3.1. Transverse HP has been selected as the force boundary condition to act on the large side of the tapered web and provides more rigidity on the edges of the hexagonal perforations. Longitudinal HP promotes stress concentration at the sharp edge corner point opposing the force direction during transmitting reciprocating cycle of the connecting rod. The web thickness of the connecting rod is 6mm. In order to have a consistent volume comparison between the equal and increasing hexagonal perforations a parameter called volume fraction (ψ) has been introduced. The volume fraction is defined as the ratio of perforation volume (ϑ_0) to the connecting rod without perforations volume (ϑ). Volume fraction for transverse hexagonal perforation expressed as equation (1) [25]

$$\psi = \frac{\vartheta_0}{\vartheta} = \frac{\left(h_1 + h_2\right)l_p t_w}{H_w L_w t_w} = \frac{\left(h_1 + h_2\right)l_p}{H_w L_w} = \frac{3}{4}\frac{h_p}{H_w}\frac{l_p}{L_w} \tag{1}$$

For Transversal Hexagonal Perforation $h_2 = 2h_1$ therefore and $h_p = \frac{4}{3}\left(h_1 + h_2\right)$

For the transverse hexagonal perforations type, the volume fractions were investigated with $\psi = 6\times10^{-2}$, $\psi = 8\times10^{-2}$, $\psi = 20\times10^{-2}$, and $\psi = 25\times10^{-2}$. The ratio h_p/l_p represents the geometric variation for each perforation. In other words, the ratio h_p/l_p promotes the perforation geometric shape variation and in order to determine critical load $P_{Critical}$ for each case, critical buckling stress are given by

$$f_{Critical} = \frac{P_{Critical}}{t_w} \tag{2}$$

3.3   REGRESSION ANALYSIS

Regression analysis refers to the study of the relationship between a response variable such as volume ratio $\left(\Psi\right)$ and one or more independent variables such as and H2/H1, hp/lp, as depicted in Figure 3.2. When the relationship follows a curve, it is called a curvilinear regression. Geometrical characteristics of hexagonal perforations are given in Table 3.2. As the volume ratio increases the web tapered ratio decreases which results in rigidity of the web of the connecting rod. The multiple regression equation in terms of volume ratio and h_1, h_2, l_p, h_p and H_2/H_1 is given in equation 3.

$$\text{Volume} - \text{Ratio}\left(\Psi\right) = -0.0247 + 0.0377\times h_1 + 0.0305\times h_2 + 0.0170\times l_p$$
$$-0.0817\times\left(H2/H1\right) \tag{3}$$

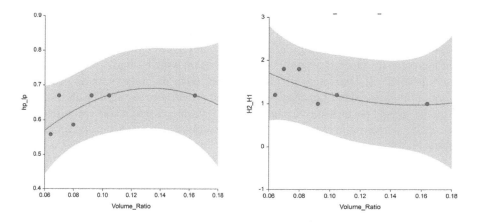

FIGURE 3.2 Variation of geometric characteristics with volume ratio: (a) hp/lp, (b) web-tapered ratio H2/H1.

TABLE 3.2
Geometrical Characteristics of Hexagonal Perforations

h1	h2	Lp	H2/H1	hp	NLS	Volume-Ratio (Ψ)
1	2	6	1	4.02	8.16362E-05	0.0921
0.5	3	7	1.2	4.69	8.14648E-05	0.1045
1.5	2	8	1.8	4.69	8.11056E-05	0.0796
0.5	3	7	1.2	4.69	8.14648E-05	0.1045
1.5	2	8	1.8	4.69	8.11056E-05	0.0796
1	2	6	1	4.02	8.16362E-05	0.0921
1.5	2	7	1.8	4.69	8.11056E-05	0.0696
1	3	8	1	5.36	8.16362E-05	0.1638
0.5	2	6	1.2	3.35	8.14648E-05	0.0640

3.4 MATERIALS AND METHODS

3.4.1 Three-Dimensional Printing Metals

3.4.1.1 Aluminum Al-Si-10Mg

Key features of Al-Si-10Mg that it has a lesser weight, good thermal conductivity, versatility in terms of hardness and strength. It is useful in rapid metal building, and has excellent machining capability. This material has many applications including complex feature dimensions, automotive, aerospace fasteners, structures, and components etc. Mechanical properties are mentioned in Table 3.3.

3.4.1.2 Stainless Steel 17-4PH

An additive manufacturing-based metal, stainless steel 17-4PH (SS17-4PH) has proven to be ideal for many industrial applications involving functional prototypes,

TABLE 3.3
Mechanical Properties and Composition of Aluminum Al-Si-10Mg

Mechanical Properties	Metric Units	Aluminum Al-Si-10Mg Composition	
Ultimate Tensile Strength	378MPa	Aluminum (Al)	Remaining %
Yield Strength	230MPa	Iron (Fe)	≤ 0.55
Modulus of Elasticity	68.4GPa	Magnesium (Mg)	≤ 0.45
Reduction of Area	10.3%	Copper (Cu)	≤ 0.05
Elongation	6.8%	Manganese (Mn)	≤ 0.45
Hardness, Rockwell B	64	Titanium (Ti)	≤ 0.15
		Silicon (Si)	9.0–11.0
		Zinc (Zn)	≤ 0.10

TABLE 3.4
Mechanical Properties and Composition of Stainless Steel 17-4PH

Mechanical Properties	Metric Units	Stainless Steel 17–4 PH Composition	
Ultimate Tensile Strength	1103MPa	Carbon (C)	0.07 Maximum
Yield Strength	1000MPa	Phosphorous (P)	0.04 Maximum
Modulus of Elasticity	197GPa	Sulfur (S)	0.03 Maximum
Reduction of Area	12%	Silicon (Si)	1.00 Maximum
Elongation	17%	Manganese (Mn)	1.00 Maximum
Hardness, Rockwell B	35	Chromium-(Cr)	15.00–17.50
		Nickel-(Ni)	3.00 to 5.00
		Niobium plus Tantalum-(Cb-Ta)	0.15–0.45
		Copper-(Cu)	3.00 to 5.00

small and medium scaled components, spare products, and mechanical fasteners. The D-M-L-S-based cost-effective prototypes offering SS17-4PH was developed with 32HRC. When SS17-4PH is heated at 900°F in an argon atmosphere its hardness reached to 42HRC. Mechanical properties are mentioned in Table 3.4.

3.4.1.3 Stainless Steel 316L

D-M-L-S-based stainless steel 316L is ideal for many applications, including consumer utensils, food processing units, chemical storage units, complex marine applications, etc. The relatively low cost of D-M-L-S stainless steel offers excellent properties for prototyping components with 85HRB. Its mechanical properties are in Table 3.5.

3.4.1.4 Titanium Ti64

Titanium Ti64 metal is characterized by very good mechanical properties and is corrosive resistant with low specific weight and biocompatibility. It is suitable for high-functional, demanding engineering applications in automotive motor racing and aerospace applications. The mechanical properties are mentioned in Table 3.6.

Advanced Manufacturing Processes

TABLE 3.5

Mechanical Properties and Composition of Stainless Steel 316L

Mechanical Properties	Metric Units	Stainless Steel 316L Composition	
Ultimate Tensile Strength	515 MPa	Carbon (C)	0.07 Maximum
Yield Strength	205 MPa	Phosphorous (P)	0.04 Maximum
Modulus of Elasticity	193GPa	Sulfur (S)	0.03 Maximum
Reduction of Area	15%	Silicon (Si)	1.00 Maximum
Elongation	45%	Manganese (Mn)	1.00 Maximum
Hardness, Rockwell B	79 HRB	Chromium (Cr)	15.00–17.50
		Nickel (Ni)	12.500–13.00
		Molybdenum-(Mo)	2.25–2.50
		Copper-(Cu)	0.50

TABLE 3.6

Mechanical Properties and Composition of Titanium Ti64

Mechanical Properties	Metric Units	Titanium Ti64 Composition	
Ultimate Tensile Strength	1210MPa	Aluminium (A)	(5.5–6.5%)
Yield Strength	880 MPa	Vanadium (V)	(3.5–4.5%)
		Oxygen (O)	< 2000 ppm
Modulus of Elasticity	950 MPa	Nitrogen (N)	< 500 ppm
Reduction of Area	36%	Carbon (C)	< 800 ppm
Elongation	14%	Hydrogen (H)	< 120 ppm
Hardness, Rockwell B	44HRC	Iron (Fe)	< 2500 ppm

3.4.2 Direct Metal Laser Sintering (D-M-L-S) Technology

A traditionally die casted aluminum, AlSi-10-Mg is usually characterized by good thermal properties, less weight, and greater strength and hardness. Aluminum AlSi-10-Mg is an economical choice as a 3DP metal for developing D-M-L-S prototypes. Prototyping with D-M-L-S-based 3D printing is the fastest, most cost effective and easiest way to turn innovative ideas into commercially marketable products. In the present decade Industry 4.0 and conventional approaches to manufacturing are enhanced their capabilities with additive manufacturing era. From removal of complexity in DFMA to developing light-weight, realistic prototypes, high-strength structures are the key features of D-M-L-S. This process, also called direct metal laser melting or laser powder bed technology, precisely develops complex geometrical features that are not feasible with other metal processing methods. D-M-L-S parts are stronger and denser in comparison to casted metal components. Additive 3D metal printing is an ideal process for intricate complex automotive, medical surgical guides, precision gas components, aerospace parts, and versatile functional 3D prototypes. D-M-L-S supplies a precision-controlled high-watt laser processing for micro-welding of powdered metals and alloys to form fully functional metal

Web Buckling Investigation

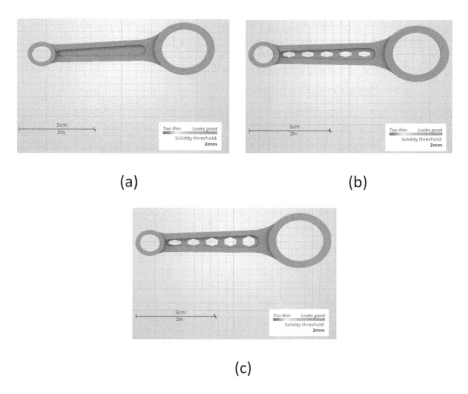

FIGURE 3.3 Solidity threshold checkup for (a) solid, (b) equal-HP, and (c) increasing-HP connecting rod before 3D metal printing.

components from CAD-based STL files. The solidity check has been done for three different designs as shown in Figure 3.3. D-M-L-S parts are developed with metals like aluminum, stainless steel, and titanium powdered materials. The components produced have an excellent surface finish in comparison to casting parts. D-M-L-S-based materials built were fully dense, had good resistance to corrosion and more robust metallic components that are treated with protective coatings. Laser sintering parameters in D-M-L-S must be selected so as to ensure sufficient re-melt of the spread layer placed and continuity of the solid-liquid metallic interface under properly controlled impingement of the laser beam layer by layer. Furthermore the blending of powders having variations in metal particle size distribution, irregular shapes, and incorrect constituent ratios will results in possibilities of porosity in powder beds, which causes heterogeneity in microstructures of D-M-L-S-based products. The solidity threshold check has been done to confirm the feasibility of metal printing, and the solidity threshold value of 2mm signifies the best density quality of metal printing of connecting rods. The molten metal flow and solidification of metal powders during single layer scanning is strongly influenced by specific laser input energy, laser scanning rates, and power. Agglomerate sizes of the D-M-L-S processed metal powders increase with increasing specific laser input energy.

3.5 CAD GEOMETRY WITH HEXAGONAL PERFORATIONS IN CONNECTING RODS

3.5.1 Criteria in the Lightweight Web Design with Hexagonal Perforations

One of the major design criteria is the stress sensitivity areas in the reduction areas and volume of connecting rod. Yield and buckling criteria have been considered in the present investigation. The web thickness of the connecting rod is constant at 6mm. The HPs have been selected as one of the methods in reducing the mass and volume of the overall weight of the connecting rod and the effects of these perforations on buckling stress have been investigated. HPs on the web have been shown in Figure 3.4.

3.5.2 Buckling Criterion

The connecting rod must oppose extreme gas pressure in the engine cylinder and compressive loads. The limit forces acting on the connecting rod depend on yield strength and its geometrical features. The perforations on the web have been given in Table 3.7. Therefore to have clear understanding of structural deformations in the connecting rod the effects of compressive buckling stress characteristics of connecting rod in terms of web section geometrical features and selected material property are taken in consideration. The compressive load of 26.0KN is applied for FE simulation on the large end section of connecting rod. The effective buckling stresses are

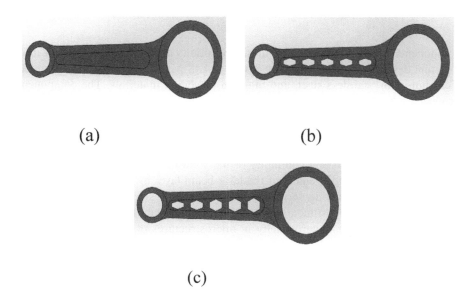

FIGURE 3.4 CAD model of connecting rod with (a) solid, (b) uniform HP, and (c) increasing HP.

TABLE 3.7
Moment of Inertia of the Web of Connecting Rod with and Without Hexagonal Perforations

Web of Connecting Rod Solid	Web of Connecting Rod Equal HP	Web of Connecting Rod Increasing HP

Web of Connecting Rod Solid

Mass = 15.19 grams

Volume = 5406.92 cubic millimeters

Surface area = 2788.75 square millimeters

Center of mass: (millimeters)
  X = -1.54
  Y = 0.00
  Z = 4.00

Principal axes of inertia and principal moments of inertia: (grams * square n
Taken at the center of mass.
  Ix = (1.00, 0.00, 0.00)   Px = 369.16
  Iy = (0.00, 1.00, 0.00)   Py = 3530.39
  Iz = (0.00, 0.00, 1.00)   Pz = 3766.31

Moments of inertia: (grams * square millimeters)
Taken at the center of mass and aligned with the output coordinate system.
  Lxx = 369.16   Lxy = 0.00   Lxz = 0.00
  Lyx = 0.00   Lyy = 3530.39   Lyz = 0.00
  Lzx = 0.00   Lzy = 0.00   Lzz = 3766.31

Moments of inertia: (grams * square millimeters)
Taken at the output coordinate system.
  Ixx = 612.26   Ixy = 0.00   Ixz = -93.70
  Iyx = 0.00   Iyy = 3809.60   Iyz = 0.00
  Izx = -93.70   Izy = 0.00   Izz = 3802.43

Web of Connecting Rod Equal HP

Mass = 4.93 grams

Volume = 4927.73 cubic millimeters

Surface area = 3216.46 square millimeters

Center of mass: (millimeters)
  X = -0.75
  Y = 0.00
  Z = 4.00

Principal axes of inertia and principal moments of inertia: (grams * square n
Taken at the center of mass.
  Ix = (1.00, 0.00, 0.00)   Px = 134.86
  Iy = (0.00, 1.00, 0.00)   Py = 1218.92
  Iz = (0.00, 0.00, 1.00)   Pz = 1308.64

Moments of inertia: (grams * square millimeters)
Taken at the center of mass and aligned with the output coordinate system.
  Lxx = 134.86   Lxy = 0.00   Lxz = 0.00
  Lyx = 0.00   Lyy = 1218.92   Lyz = 0.00
  Lzx = 0.00   Lzy = 0.00   Lzz = 1308.64

Moments of inertia: (grams * square millimeters)
Taken at the output coordinate system.
  Ixx = 213.70   Ixy = 0.00   Ixz = -14.80
  Iyx = 0.00   Iyy = 1300.55   Iyz = 0.00
  Izx = -14.80   Izy = 0.00   Izz = 1311.41

Web of Connecting Rod Increasing HP

Mass = 4.59 grams

Volume = 4588.56 cubic millimeters

Surface area = 3158.60 square millimeters

Center of mass: (millimeters)
  X = -1.54
  Y = 0.00
  Z = 4.00

Principal axes of inertia and principal moments of inertia: (grams * square n
Taken at the center of mass.
  Ix = (1.00, 0.00, 0.00)   Px = 132.16
  Iy = (0.00, 1.00, 0.00)   Py = 1128.82
  Iz = (0.00, 0.00, 1.00)   Pz = 1217.87

Moments of inertia: (grams * square millimeters)
Taken at the center of mass and aligned with the output coordinate system.
  Lxx = 132.16   Lxy = 0.00   Lxz = 0.00
  Lyx = 0.00   Lyy = 1128.82   Lyz = 0.00
  Lzx = 0.00   Lzy = 0.00   Lzz = 1217.87

Moments of inertia: (grams * square millimeters)
Taken at the output coordinate system.
  Ixx = 205.57   Ixy = 0.00   Ixz = -28.25
  Iyx = 0.00   Iyy = 1213.11   Iyz = 0.00
  Izx = -28.25   Izy = 0.00   Izz = 1228.74

TABLE 3.8
Moment of Inertia of the Web of Connecting Rod with and Without Hexagonal Perforations

Solid Connecting rod	Connecting rod Equal HP	Connecting rod Increasing HP

Mass = 48.04 grams	Mass = 16.38 grams	Mass = 16.04 grams
Volume = 17095.78 cubic millimeters	Volume = 16383.12 cubic millimeters	Volume = 16043.94 cubic millimeters
Surface area = 9017.09 square millimeters	Surface area = 9379.54 square millimeters	Surface area = 9345.68 square millimeters
Center of mass: (millimeters) X = 21.30 Y = 0.00 Z = 4.00	Center of mass: (millimeters) X = 22.28 Y = 0.00 Z = 4.00	Center of mass: (millimeters) X = 22.55 Y = 0.00 Z = 4.00
Principal axes of inertia and principal moments of inertia: (grams * square n Taken at the center of mass. Ix = (1.00, 0.00, 0.00) Px = 4577.77 Iy = (0.00, 1.00, 0.00) Py = 63921.30 Iz = (0.00, 0.00, 1.00) Pz = 67419.22	Principal axes of inertia and principal moments of inertia: (grams * square n Taken at the center of mass. Ix = (1.00, 0.00, 0.00) Px = 1626.33 Iy = (0.00, 1.00, 0.00) Py = 22162.91 Iz = (0.00, 0.00, 1.00) Pz = 23409.23	Principal axes of inertia and principal moments of inertia: (grams * square n Taken at the center of mass. Ix = (1.00, 0.00, 0.00) Px = 1623.63 Iy = (0.00, 1.00, 0.00) Py = 22061.19 Iz = (0.00, 0.00, 1.00) Pz = 23306.84
Moments of inertia: (grams * square millimeters) Taken at the center of mass and aligned with the output coordinate system. Lxx = 4577.77 Lxy = 0.02 Lxz = 0.00 Lyx = 0.02 Lyy = 63921.30 Lyz = 0.00 Lzx = 0.00 Lzy = 0.00 Lzz = 67419.22	Moments of inertia: (grams * square millimeters) Taken at the center of mass and aligned with the output coordinate system. Lxx = 1626.33 Lxy = 0.01 Lxz = 0.00 Lyx = 0.01 Lyy = 22162.91 Lyz = 0.00 Lzx = 0.00 Lzy = 0.00 Lzz = 23409.23	Moments of inertia: (grams * square millimeters) Taken at the center of mass and aligned with the output coordinate system. Lxx = 1623.63 Lxy = 0.01 Lxz = 0.00 Lyx = 0.01 Lyy = 22061.19 Lyz = 0.00 Lzx = 0.00 Lzy = 0.00 Lzz = 23306.84
Moments of inertia: (grams * square millimeters) Taken at the output coordinate system. Ixx = 5346.39 Ixy = 0.02 Ixz = 4091.99 Iyx = 0.02 Iyy = 86474.71 Iyz = 0.00 Izx = 4091.99 Izy = 0.00 Izz = 89204.00	Moments of inertia: (grams * square millimeters) Taken at the output coordinate system. Ixx = 1888.46 Ixy = 0.01 Ixz = 1460.36 Iyx = 0.01 Iyy = 30560.87 Iyz = 0.00 Izx = 1460.36 Izy = 0.00 Izz = 31545.05	Moments of inertia: (grams * square millimeters) Taken at the output coordinate system. Ixx = 1880.33 Ixy = 0.01 Ixz = 1446.91 Iyx = 0.01 Iyy = 30473.43 Iyz = 0.00 Izx = 1446.91 Izy = 0.00 Izz = 31462.37

Web Buckling Investigation

calculated at the part of web section using ultimate strength of each of the 3DP metal properties. Normal limiting stress (NLS) for buckling is defined as

$$\text{NLS} = \frac{f_{critical}}{f_y} \tag{4}$$

Finally, using this equation for different 3DP metals, the geometric characteristics of the proposed hexagonal perforations are compared. To analyze the buckling behavior, the Finite element analysis (FEA) approach was used for predicting buckling stresses with different perforations as given in Table 3.8.

3.6 RESULT AND DISCUSSIONS

3.6.1 STRUCTURAL ANALYSIS

The structural analysis has been carried out with FE-based analysis by using the SIMSOLID tool, and the results are tabulated in Table 3.9. It has been observed from the results that displacement magnitude is less in cases of connecting rods with equal HP in SS-17-4 at peak frequency as shown in Figure 3.5.

3.6.2 MODAL ANALYSIS AND FREQUENCY RESPONSE

The dynamic frequency response of all the connecting rods using different 3DP metals has been analyzed and their ten mode frequencies have been obtained. The fundamental frequency of the solid connecting rod is 8.4860e+02, which is higher in comparison with equal HP and increasing HP. The fundamental frequency of the increasing HP connecting rod is found to be less and it provides more rigidity during the action of buckling load. The modal frequencies are shown in Figures 3.6, 3.7, and 3.8. It can be observed from Table 3.10 that higher modes of frequencies are lesser

TABLE 3.9

Displacement at Peak Frequency of Connecting Rods

	Connecting Rod		
	Solid	Equal HP	Increasing HP
	AlSiMg0.6 (At Peak Frequency)		
Displacement Magnitude, mm	5.7508	6.4344	5.5867
	Stainless-Steel-17-4 (At Peak Frequency)		
Displacement Magnitude, mm	1.3722	2.1476	1.9901
	Stainless Steel 316L (At Peak Frequency)		
Displacement Magnitude, mm	1.1430	2.3764	1.7584
	Titanium-Ti-Al-6 (At Peak Frequency)		
Displacement Magnitude, mm	3.3958	3.8528	4.2574

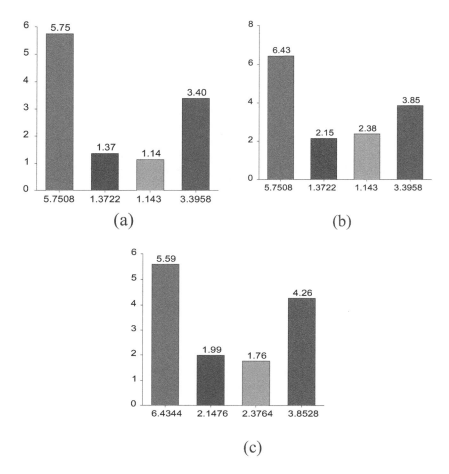

FIGURE 3.5 Deformation of the connecting rod with (a) solid (b) equal HP, and (c) increasing HP.

when compared to fundamental frequencies. Therefore the D-M-L-S-based connecting rods proved to be suitable for buckling load applications in automobile engines.

A minimum buckling moment in the case of equal HP has been observed for AlSiMg0.6 among different hexagonal perforations and 3DP metals as shown in Figure 3.9. The buckling moment is least for stainless steel 316L among all 3DP metallic connecting rods. The buckling moments of the 3DP metals are given in Table 3.11.

3.6.3 Buckling Analysis

The displacement magnitude and von Mises stress have been shown in Figures 3.10, 3.11, and 3.12 for design 1, 2, and 3 for different buckling modes as stated earlier. The normal limiting stress and characteristics ratios hp/lp are determined by factors

Web Buckling Investigation

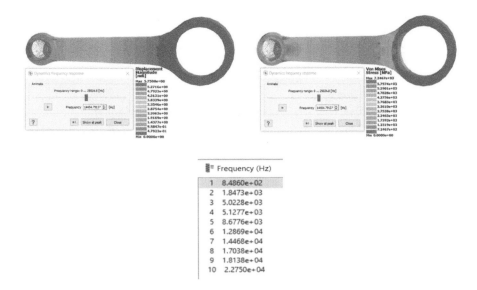

FIGURE 3.6 DFR of solid aluminum AlSiMg0.6 of connecting rods.

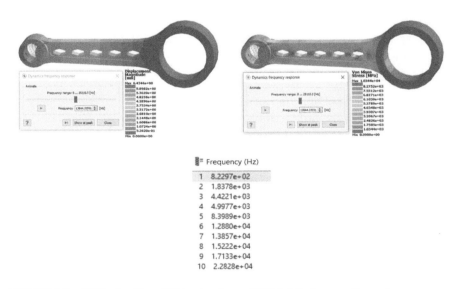

FIGURE 3.7 DFR of uniform HP in aluminum AlSiMg0.6 of connecting rods.

such as the elastic modulus of different 3DP metals with respect to tapered ratio H2/H1, perforation characteristics ratio hp/lp, volume ratio, and buckling load conditions. The FE-based stress data has been given in Table 3.12, and the displacement magnitude was observed to be less in comparison with other HPs. Designers should consider the weight optimization criteria and enhancement of overall efficiency of

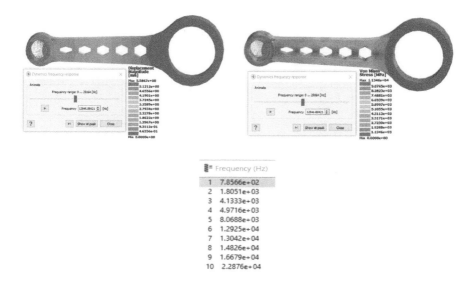

FIGURE 3.8 DFR of increasing HP in aluminum AlSiMg0.6 of connecting rods.

TABLE 3.10
Modal Frequencies of Connecting Rods with Different Hexagonal Perforations

Modes	Connecting rod		
	Solid	Equal HP	Increasing HP
1.	8.4860e+02	8.2297e+02	7.8566e+02
2.	1.8473e+03	1.8378e+03	1.8051e+03
3.	5.0228e+03	4.4221e+03	4.1333e+3
4.	5.1277e+03	4.9977e+03	4.9716e+03
5.	8.6776e+03	8.3989e+03	8.0688e+03
6.	1.2869e+04	1.2880e+04	1.2925e+04
7.	1.4468e+04	1.3857e+04	1.3042e+04
8.	1.7038e+04	1.5222e+04	1.4826e+04
9.	1.8138e+04	1.7133e+04	1.6679e+04
10.	2.2750e+04	2.2828e+04	2.2876e+04

the engine by considering equal HP as the optimum geometric characteristic of HP for design and development of connecting rods.

3.6.4 Variation of Normalized Limiting Stress with Geometric Characteristics

3.6.4.1 AlSiMg0.6

Contour plots are topographical regions obtained from 3D data points and constructed from three geometric characteristics. The geometric characteristics such as

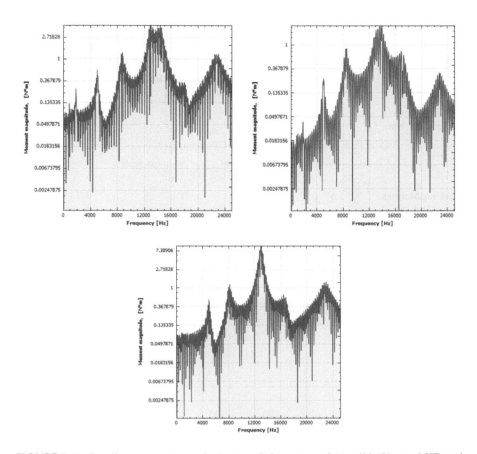

FIGURE 3.9 Bending moment magnitude at peak frequency of (a) solid, (b) equal HP, and (c) increasing HP in AlSiMg0.6 connecting rods.

HP are represented on the horizontal axis and lp is represented on the vertical axis. The third variable, NLS, is represented by a color gradient and isolines. These contour plots are often useful in data analysis in determining maximums and minimums from the set of trivariate data. Figure 3.13 are presents the variation of the NLS versus taper ratio H2/H1. It can be seen in Figure 3.13b that NLS decreases with the increase of the tapered ratio as it is confirming with the fact that the increase in slenderness ratio leads to consistent decrease of NLS. The graphical representation of NLS in Figure 3.14a shows that there is marginal increase of the NLS as volume ratio increases, as in the case of increasing hexagonal perforations. The range of variable between 5.0<hp<6.0 and 7.8<lp<8.0, it has been observed from Figure 3.14b, there will be increase in NLS values for Al-Si-10Mg.

3.6.4.2 Stainless Steel 17_4PH

The curve fitting has been done by using the parameters of the Michaelis-Menten equation. The parameters included in this model represent both graphics and numerical tests data such as F-test for curve coincidence and compares curve coincidence

TABLE 3.11
Buckling Moment Magnitude at Peak Frequency of Solid Connecting Rods

	AlSiMg0.6	Stainless-Steel-17-4	Stainless Steel 316L	Titanium-Ti-Al-6
Buckling Moment Magnitude, Nm	2.8	1.6	1.7	2.9

Moment Magnitude at Peak Frequency of Equal HP in Connecting Rod

	AlSiMg0.6	Stainless-Steel-17-4	Stainless Steel 316L	Titanium-Ti-Al-6
Buckling Moment Magnitude, Nm	1.6	1.7	1.4	1.7

	AlSiMg0.6	Stainless-Steel-17-4	Stainless Steel 316L	Titanium-Ti-Al-6
Buckling Moment Magnitude, Nm	1.6	1.7	1.4	1.7

Moment Magnitude at Peak Frequency of Increasing HP in Connecting Rod

	AlSiMg0.6	Stainless-Steel-17-4	Stainless Steel 316L	Titanium-Ti-Al-6
Buckling Moment Magnitude, Nm	7.5	7.6	7.5	7.6

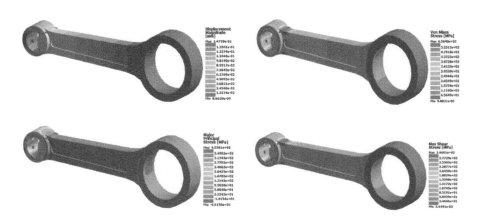

FIGURE 3.10 Design of case 1: without hexagonal perforations in web of connecting rod.

limit with desired geometric characteristics values. Contour plots are topographical regions obtained from 3D data points and constructed from three geometric characteristics. The geometric characteristics such as hp are represented on the horizontal axis and lp is represented on the vertical axis. The third variable NLS is represented by a color gradient and isolines. These contour plots are often useful in data analysis in

Web Buckling Investigation

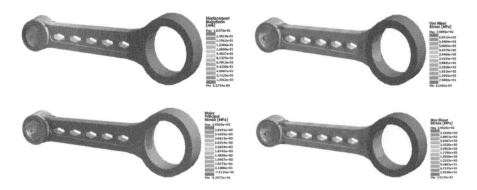

FIGURE 3.11 Design of case 2: equal hexagonal perforations in web of connecting rod.

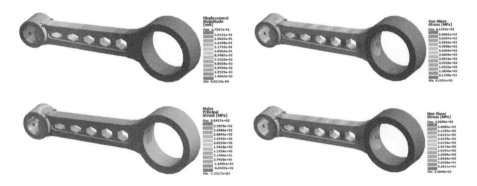

FIGURE 3.12 Design of case 3: increasing hexagonal perforations in web of connecting rods.

TABLE 3.12
FEA-based Stress Data

	Equal HP	Increasing HP	Solid
Displacement Magnitude, mm	1.627e-01	1.759e-01	1.472e-01
von Mises Stress, MPa	7.589e+02	8.135e+02	6.564e+02
Maximum Principal Stress, MPa	4.930e+02	5.042e+02	4.538e+02
Maximum Principal Strain	2.453e-03	2.454e-03	2.282e-03
Energy Density, MPa	1.546e+00	1.746e+00	1.189e+00
Maximum Shear Stress, MPa	3.952e+02	4.260e+02	3.466e+02

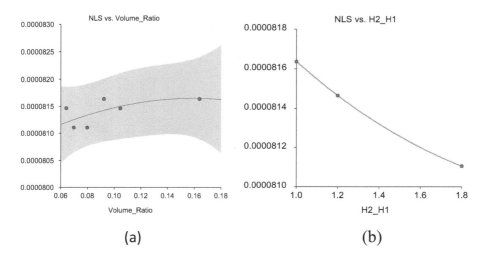

FIGURE 3.13 Variation of NLS of Al_Si_10Mg (a) volume ratio and (b) web tapered ratio H2/H1.

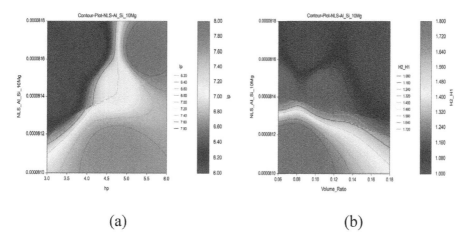

FIGURE 3.14 Variation of NLS of Al_Si_10Mg (a) hp and lp and (b) volume Ratio and tapered ratio H2/H1.

determining maximums and minimums from the set of trivariate data. Figure 3.15a, represent the variation of the Normal Limiting Stress versus volume ratio. It has been observed from the Figure 3.15b that NLS decreases with the increase of the tapered ratio as it is confirming with the fact that the increase in slenderness ratio leads to consistent decrease of NLS. The graphical representation of NLS in Figure 3.16a shows that there is marginal increase of the NLS as the increase of volume ratio as in case of increasing hexagonal perforations. The similar trend has been observed in the range of hp between 5.0<hp<6.0 and 7.6<lp<8.0, and it is also observed from Figure 3.16b that there is an increase in NLS values.

Web Buckling Investigation

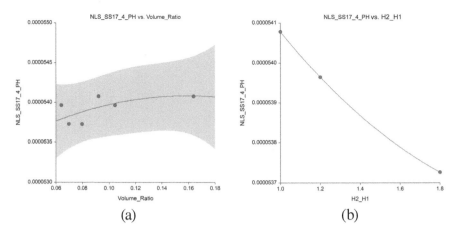

FIGURE 3.15 Variation of NLS of stainless steel 17_4PH (a) volume ratio and (b) web tapered ratio H2/H1.

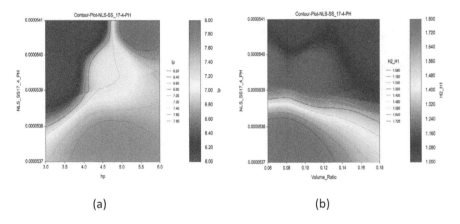

FIGURE 3.16 Variation of NLS of Stainless Steel 17_4PH (a) hp and lp and (b) volume and tapered ratio H2/H1.

3.6.4.3 Stainless Steel 316L

A similar trend has been observed in case of stainless steel 316L for the values of NLS with respect to volume ratio and tapered ratio H2/H1. Figure 3.17a represents the marginal increase of the NLS with volume ratio, and Figure 3.17b shows that NLS decreases with the increase of the tapered ratio H2/H1. The graphical representation of NLS in Figures 3.18a and 3.18b shows that there is marginal increase of the NLS as volume ratio increases, as in the case of increasing HPs.

3.6.4.4 Titanium Ti64

The values of NLS with respect to volume ratio and tapered ratio H2/H1 for titanium Ti64 has been shown in Figure 3.19a and represent the marginal increase of the NLS from the initial value 0.000047575 with 0.18 as volume ratio. It can be observed from

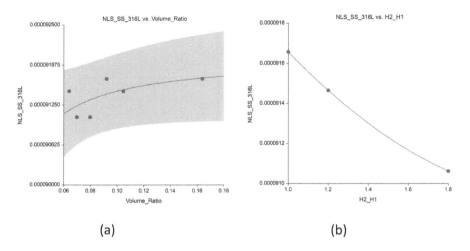

FIGURE 3.17 Variation of NLS of stainless steel 316L (a) volume ratio and (b) web tapered ratio H2/H1.

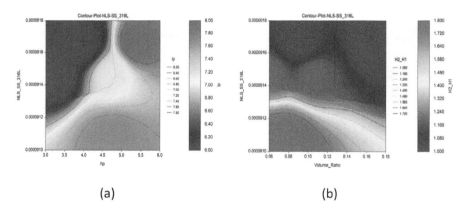

FIGURE 3.18 Variation of NLS of stainless steel 316L (c) hp and lp and (d) volume ratio and tapered ratio H2/H1.

the Figure 3.19b that NLS decreases with the increase of the tapered ratio H2/H1. The graphical representation of NLS in Figure 3.20a and 3.20b shows that there is marginal increase of the NLS as the volume ratio increases, as in case of increasing HPs.

Surface plots are the graphical calculators that help the designers in selecting the proper geometric characteristic parameters in the 3D data as shown in Figure 3.21. The dark black region of the surface plots are the desired region of optimum dimensions of the HP for the connecting rod. The surface plots surface is constructed using the averaged values of tapered ratio H2/H1. These surface plots are useful in development of regression analysis for viewing the relationship among a dependent and two independent variables.

Web Buckling Investigation

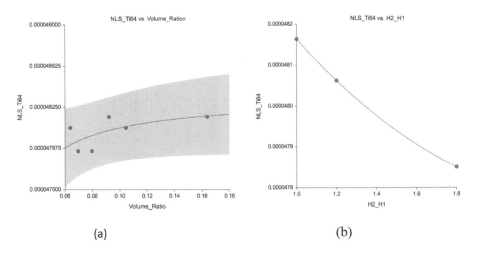

FIGURE 3.19 Variation of NLS of titanium Ti64 (a) volume ratio and (b) web tapered ratio H2/H1.

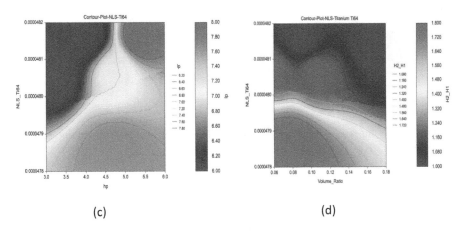

FIGURE 3.20 Variation of NLS of titanium Ti64 (c) hp and lp and (d) volume ration and tapered ratio H2/H1.

3.7 CONCLUSIONS

The resultant stress data was obtained for the D-M-L-S-based connecting rod by using finite element procedure of SIM-SOLID-20.0. The four different 3DP metals were selected for the connecting rod, namely, AlSiMg0.6, Stainless-Steel-17-4, Stainless Steel 316L, and Titanium-Ti-Al-6. Structural analyses including displacement, max principal stress, max principal strain, energy density, von Mises, and shear stress have been calculated for different designs for the solid connecting. On comparing these FE-based results, it is noted that stainless steel 316L material gives

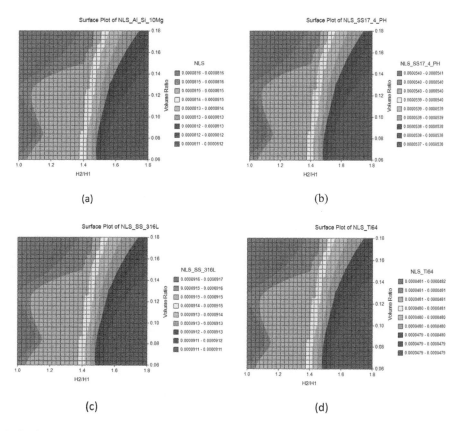

FIGURE 3.21 Surface plots of variations in different 3DP metals with volume ratio and tapered ratio H2/H1.

least total deformation, equivalent elastic strain, and equivalent stress among all the tested materials. The displacement of stainless-steel-17-4 with equal HP is minimum when compared to other printing metals at peak frequency. The percentage decrease in volume of connecting rod having equal hexagonal perforations is 4.16 percent in comparison with solid connecting rod and similarly 6.15 percent decrease in volume in comparison with increasing hexagonal perforations. Von Mises stress is observed to be 10.53 percent increase in equal HP when compare to solid connecting rods, but the overall weight reduction to 65.90 percent leads to increase the fuel efficiency due to less rotational masses. On the other hand, stainless steel 316L has the minimum buckling moment and equivalent stress distribution in comparison to different design configurations. The solid connecting rod has less deformation by considering D-M-L-S-based metal printing but it cannot contribute to enhancement of the engine fuel efficiency. It can be further concluded from the FE simulation data that design of Equal HP based 3DP connecting rods should be considered as a design for developing the connecting rod by D-M-L-S. However the limitations of producing the connecting rod economically need focused controlling parameters based upon the number of parts producing capacity of the machine. Further extensions of the present work in the future may include development of connecting rods at different sintering temperatures, scanning power, powder size, and flow rate of the molten metal and experiments to evaluate web buckling behaviors of connecting rods.

Nomenclature:

3DP	Three-Dimensional Printing
D-M-L-S	Direct Metal Laser Sintering
HP	Hexagonal Perforations
NLS	Normal Limiting Stress
hp	Width of Hexagonal Perforation
lp	Length of Hexagonal Perforation
ψ	Volume Ratio
H2/H1	Tapered Ratio
H_w	Height of Web of I section
L_w	Length of the Web
t_w	Thickness of the Web
h_1	Geometric Characteristics of Hexagons
h_2	Geometric Characteristics of Hexagons
$P_{Critical}$	Critical Buckling Load
$f_{critical}$	Critical Buckling Stress
f_y	Yield Strength of D-M-L-S-based metals

3.8 ACKNOWLEDGEMENT

The authors would like to thank the Center of Excellence for Advanced Manufacturing Engineering, Mechanical Design and Manufacturing Engineering (MDME), School of Mechanical, Chemical and Materials Engineering (SoMCME), ADAMA Science and Technology University (ASTU), Ethiopia.

3.9 CONFLICTS OF INTEREST

The main corresponding author states that there is no conflict of interest. The authors also declare that they have no financial assistance from internal or external funding bodies.

REFERENCES

[1] A.A.R.M. Abad, M. Ranjbarkohan, and B.N. Dardashti, "Dynamic load analysis and optimization of connecting rod of samand engine," *Austr. J. Basic Appl. Sci.*, vol. 5, no. 12, pp. 1830–1838, 2011.

[2] S. Kunal and S. AkhandPratap, "A review paper on design analysis of internal combustion components," *Int. J. Adv. Res. Sci. Eng.*, vol. 6, pp. 495–498, 2017.

[3] M.K. Lee, H. Lee, T. Lee, and H. Jang, "Buckling sensitivity of a connecting rod to the shank sectional area reduction," *J. Mater. Des.*, vol. 31, pp. 2796–2803, 2004.

[4] S. Kodiyalam, G. N. Vanderplaats, and H. Miura, "Structural shape optimization with MSC/Nastran," *Comput. Struct.*, vol. 40, no. 4, pp. 821–829, 1991.

[5] D. Gopinath, and C.V. Sushma, "Design and optimization of four wheeler connecting rod using finite element analysis," *4th Int. Conf. Mater. Proc. Char. Materials Today: Proc.*, vol. 2, pp. 2291–2299, 2015.

[6] M. E. M. El-Sayed and E. H. Lund, "Structural optimization with fatigue life constraints," *Eng. Fract. Mech.*, vol. 37, no. 6, pp. 1149–1156, 1990.

[7] R. Hippliti, "FEM method for design and optimization of connecting rods for small two-stroke engines," *Small Eng. Technol. Conf.*, pp. 217–231, 1993.

[8] P. C. Lu, "The shape optimization of a connecting rod with fatigue life constraint," *Int. J. Mater. Prod. Technol.*, vol. 11, no. 5–6, pp. 357–370, 1996.

[9] B. Talikoti, S. N. Kurbet, V. V. Kuppast, and M. Arvind, "Harmonic analysis of a two-cylinder crankshaft using ANSYS," 2016 International Conference on Inventive Computation Technologies (ICICT), Coimbatore, India, 2016.

[10] Z. Q. Zhang, J. X. Zhou, N. Zhou, X. M. Wang, and L. Zhang, "Shape optimization using reproducing kernel particle method and an enriched genetic algorithm," *Comput. Methods Appl. Mech. Eng.*, vol. 194, pp. 4048–4070, 2005.

[11] T. H. Lee andJ. J. Jung, "Metamodel-based shape optimization of connecting rod considering fatigue life," *Key Eng Mater*, vol. 306–308, no. 1, pp. 211–216, 2006;P. S. Shenoy and A. Fatemi, "Connecting rod optimization for weight and cost reduction," *SAE Technical Paper 2005–01–0987*, 2005.

[12] Z. Honggen, L. Shan, L. Guochao, T. Guizhong, W. Ziyu, and W. Chuhui, "Machining stress analysis and deformation prediction of connecting rod based on FEM and GRNN," *Iranian J. Sci. Technol. Trans. Mech. Eng.*, vol. 44, pp. 183–192, 2020.

[13] LucjanWiteka Paweł Zelek, "Stress and failure analysis of the connecting rod of diesel engine," *Eng. Fail. Anal.*, vol. 97, pp. 374–382, 2019.

[14] J. H. Martin, B. D. Yahata, J. M. Hundley, J. A. Mayer, T. A. Schaedler, and T. M. Pollock, "3D printing of high-strength aluminium alloys," *Nature.*, vol. 549, no. 7672, pp. 365–369, 2017. doi:10.1038/nature23894

[15] D. Zhang, D. Qiu, M. A. Gibson, et al., "Additive manufacturing of ultrafine-grained high-strength titanium alloys [published correction appears in Nature. 2020 June;582(7811):E5]," *Nature.*, vol. 576, no. 7785, pp. 91–95, 2019. doi:10.1038/s41586-019-1783-1

[16] F. Trevisan, F. Calignano, M. Lorusso, et al., "On the selective laser melting (SLM) of the AlSi10Mg Alloy: Process, microstructure, and mechanical properties," *Materials (Basel).*, vol. 10, no. 1, p. 76, 2017. Published 2017 Jan 18. doi:10.3390/ma10010076

[17] S. Ford and M. Despeisse, "Additive manufacturing and sustainability: An exploratory study of the advantages and challenges," *J. Clean. Prod.*, vol. 137, pp. 1573–1587, 2016.

[18] L. E. Murr, "A metallographic review of 3D printing/additive manufacturing of metal and alloy products and components," Published online: 12 March 2018 Springer Science & Business Media, LLC, part of Springer Nature and ASM International 2018.

[19] www.stratasysdirect.com/technologies/direct-metal-laser-sintering.

[20] G. Mohammed, S. Ahmed, and S. Algarni, "Design, development and FE thermal analysis of a radially grooved brake disc developed through direct metal laser sintering," *Materials*, vol. 11, no. 7, p. 1211, 2018. https://doi.org/10.3390/ma11071211

[21] B. G. Compton and J. A. Lewis, "3d-printing of lightweight cellular composites," *Adv. Mat.*, vol. 26, no. 34, pp. 5930–5935, 2014.

[22] C. B. Williams, J. K. Cochran, and D. W. Rosen, "Additive manufacturing of metallic cellular materials via three-dimensional printing," *Int. J. Adv. Manuf. Techn.*, vol. 53, no. 1, pp. 231–239, 2011.

[23] C. Rameshbabu and S. Prabavath, "Simplified design equation for web tapered—Isections using finite element modeling," https://doi.org/10.2478/ace-2018-0029, Published online: 22 May 2019.

[24] Anisio Andrade et al., "Lateral torsional buckling of singly symmetric web—tapered thin walled I beams: 1D Model vs Shell FEA," *Comput. Struct.*, vol. 85, pp. 1343–1339, 2007.

[25] Caio César Cardoso da Silva, Daniel Helbig, Marcelo Langhinrichs Cunha, Luiz Alberto Oliveira Rocha, Elizaldo Domingues dos Santos, Mauro-de Vasconcellos Real, Liércio André Isoldi, Numerical buckling analysis of thin steel plates with centered hexagonal perforation through constructal design method Received: 31 December 2018/ Accepted: 27 June 2019 © The Brazilian Society of Mechanical Sciences and Engineering 2019.

[26] G. Lorenzini, D. Helbig, C. C. C. Silva, M. V. Real, E. D. Santos, L. A. Isoldi, and L. A. O. Rocha, "Numerical evaluation of the effect of type and shape of perforations on the buckling of thin steel plates by means of the constructal design method," *Int. J. Heat Technol.*, vol. 34, pp. 9–20, 2016.

[27] S. Kwani and P. K. Wijaya, "Lateral torsional buckling of castellated beams analyzed using the collapse analysis," *Procedia Eng.*, vol. 171, pp. 813–820, 2017.

[28] W. Yuan, N. Yu, Z. Bao, and L. Wu, "Defection of castellated beams subjected to uniformly distributed transverse load," *Int. J. Steel Struct.*, vol. 16, no. 3, pp. 813–821, 2016.

[29] M.R. Soltani, A. Bouchair, and F. E. Nonlinear, "Analysis of the ultimate behavior of steel castellated beams," *J. Sci. Common.*, vol. 70, pp. 101–114, 2012.

[30] P. Zhang, J. Toman, Y. Yu, E. Biyikli, M. Kirca, M. Chmielus, and A. C. To, "Efficient design optimization of variable-density hexagonal cellular structure by additive manufacturing: Theory and validation," *J. Manuf. Sci. Eng.*, vol. 137, no. 2, p. 021004, 2015.

[31] www.altair.com, www.altair.com/trysimsolid.

[32] P. K. Chaurasiya, N. K. Sahu, R. Dwivedi, S. K. Singh, and S. Chhalotre, "Buckling sensitivity of ultrafine grained material AA 2618 connecting rod," *Int. J. Recent Technol. Eng.*, vol. 8, no. 6, 2020.

[33] G. Partskhaladze, I. Mshevenieradze, E. Medzmariashvili, G. Chaleshvili, V. Yepes, and J. Alcala, "Buckling Analysis and stability of low carbon steel rods in the elastoplastic region of materials," *Adv. Civ. Eng.*, Hindawi, vol. 2019, 2019, ID 7601260.

4 Materials for Additive Manufacturing

Concept, Technologies, Applications and Advancements

M. Anugrahaprada and Pawan Sharma

4.1	Introduction	79
4.2	Additive Manufacturing Processes	80
	4.2.1 Fused Deposition Modelling (FDM)	81
	4.2.2 Selective Laser Sintering (SLS)	82
	4.2.3 Stereolithography (SLA)	82
4.3	Additive Manufacturing Materials	83
	4.3.1 Metals	83
	4.3.2 Ceramics	86
	4.3.3 Polymers	88
4.4	Applications	90
	4.4.1 Aerospace	90
	4.4.2 Biomaterials	91
4.5	Future Scope	92
4.6	Conclusion	93
References		94

4.1 INTRODUCTION

Additive manufacturing (AM) is described as a process of joining materials to build solid parts from 3D model data, one layer upon another [1]. In contrast to machining, stamping and other conventional manufacturing techniques where the part is fabricated by material removal from more extensive sheet metal, AM produces the desired shape by material addition by efficiently using raw materials, producing minimal waste and giving high-dimensional accuracy [2]. AM involves a variety of techniques, materials and apparatus and has been developed over the years with the capability of evolving manufacturing and logistics processes [3].

Modern AM or 3D printing dates back to the mid-1980s. Charles Hull, in the year 1986, developed this technology through stereolithography (SLA). It was succeeded

DOI: 10.1201/9781003220237-4

by consequent progress such as inkjet printing, fused deposition modelling (FDM), contour crafting (CC) and powder bed fusion [3]. The SLA-1, the first commercial AM fabricator, was sold in 1988 by 3D Systems [4]. Patents were granted in 1986, following which different companies like Cubital (Solid Ground Curing, SGC), Helisys (Laminated Object Manufacture or LOM) and DTM (Selective Laser Sintering, SLS), respectively employed the SLS technique in respective applications. Scott Crump in 1989 patented fused deposition modelling (FDM) and formed the Stratasys Company. The 3D printing (3DP) process was also patented in 1989 by a group from MIT [5]. Several other commercially successful systems were developed in the 1980s and 1990s. In 2009, ASTM classified AM processes into seven methodologies [1]:

   i. Material extrusion
   ii. Sheet lamination
  iii. Vat polymerization
  iv. Material jetting
   v. Powder bed fusion
  vi. Binder jetting
 vii. Directed energy deposition

Further, the advent of computer geometric modelling and the introduction of personal computers led to AM gaining immense popularity in the world as this technology became accessible. Today, AM finds commercial applications in the biomedical, automotive and aerospace industries, architecture, buildings and protective structures. However, AM also has its limitations. It is difficult and time consuming to produce large sized parts due to poor strength of material. AM parts sometimes might have a rough and unfinished appearance due to the stacking of large powder particles of the raw material. AM is also an expensive technology since the cost of AM equipment and materials is high. Research is being carried out to overcome these limitations [2].

4.2 ADDITIVE MANUFACTURING PROCESSES

The generic AM process involves three necessary steps [2]:

   i. Development of a computerized 3D model of the required part and its conversion to a prescribed format for AM files, for example, the conventional standard tessellation language format (STL) or other modern AM file formats.
  ii. Sending the file to an AM machine where the solid model undergoes a change in its position and orientation is scaled or manipulated in other ways.
 iii. Layer by layer building of the part on the AM machine.

Various AM methods create and integrate layers in several ways. For instance, some methods utilize thermal energy from laser or electron beams which aid in melting or sintering metal or plastic powder jointly. In contrast, other methods make use of inkjet printing heads to precisely spray binders or solvents onto polymer or ceramic

Materials for Additive Manufacturing

powders [2]. Some commonly used AM processes are described in subsequent sections.

4.2.1 Fused Deposition Modelling (FDM)

FDM is a relatively recent AM process developed in the 1980s [6]. Two types of materials are generally used as feedstock materials to construct the part; the support material and model material, both supplied as wires wound on spools. These wires are then fed inside the machine's head, where the material is heated to a partial liquid condition. The heating process in the head is temperature-controlled. Then, the material is extruded from the machine head, depositing on the fixtureless base of the FDM machine. The newly extruded layers keep bonding with the previous layers and solidify. The desired 3D solid part gets built in this manner. The support structure is then removed after the fabrication of the part. Figure 4.1 depicts the schematic diagram of FDM.

The FDM machine is enabled to operate in the X, Y and Z axes. Typically, the layer thickness in FDM varies from 0.127 to 0.254 mm [7]. This process is advantageous as myriads of materials have the potential to be utilized. The accuracy of the finished product achieved is around ±0.05 mm. The equipment and maintenance cost for FDM is also low. However, the seam line between layers, the requirement of support, longer build time, and delamination resulting from temperature fluctuations pose challenges to the FDM process [2].

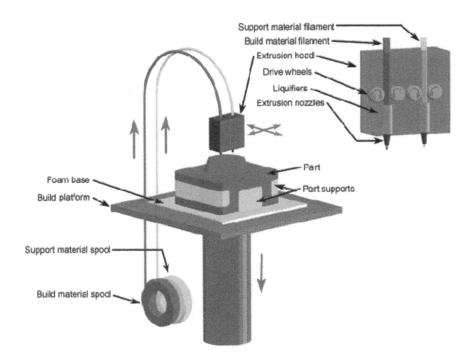

FIGURE 4.1 Schematic diagram of FDM [6].

4.2.2 Selective Laser Sintering (SLS)

SLS was first built by The University of Texas in Austin and patented in 1989. DTM Corporation commercially pursued it outside of Austin [8–9]. It is a process of generating solid models and a wide range of materials to do so. As powders, the raw material is deposited in layers fused using laser light, which forms the desired products. There are three chambers in the SLS system; the central chamber or part bed where the model is built and the two other chambers where the powder is stocked before its supply across the build area, employing a roller [9]. A piston is used to adjust the powder layer by moving downward by the one-layer thickness, while the extra powder provides support to the part when being built. A laser beam traces over the powder surface, its heat selectively melting the powder as the laser strikes it under the scanning system's control. The laser's heat is necessary to elevate the temperature. The central chamber is kept at a temperature just beneath the powder material's melting point, which slightly causes sintering [8]. Figure 4.2 shows the schematic diagram of SLS.

SLS finds vast applications in various industries. The powder used as the raw material can be recycled without significant changes in its mechanical properties. However, sub-par surface finish, poor dimensional accuracy, and microstructural and mechanical properties seldom matching industrial requirements are some SLS limitations [10].

4.2.3 Stereolithography (SLA)

SLA, developed in 1986 by 3D Systems, is the first commercial solid free form (SFF) technique [12]. SLA incorporates curing or solidifying photosensitive polymers in the liquid state by employing an irradiation source of light. This provides the energy needed to cause a chemical reaction, i.e., curing, leading to bond formation between

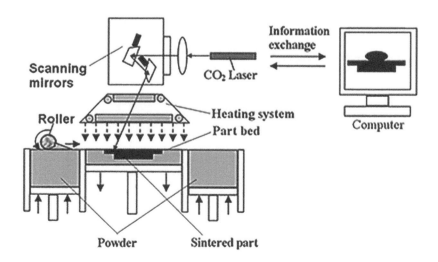

FIGURE 4.2 Schematic diagram of SLS [11].

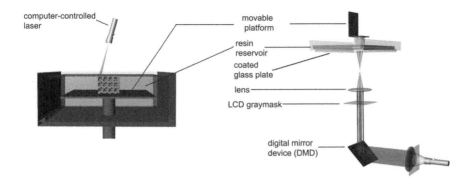

FIGURE 4.3 Schematic diagram of SLA [12].

myriads of tiny molecules. Consequently, a highly cross-linked polymer is formed [13]. This reaction is an exothermic polymerization reaction known as photo polymerization, during which two transitions, namely gelation and vitrification, occur. During gelation, a liquid-to-rubber transition takes place, which leads to a significant rise in viscosity. Vitrification is the thermodynamically reversible transition from liquid or rubber resins to glassy and solid resins [14]. The resin solidification process takes place on a platform. The prescribed depth for curing is somewhat greater than the platform's step height, ensuring significant bonding with the first layer. The curing and solidification process is continued until the 3D solid product is obtained. Figure 4.3 represents the schematic diagram of SLA.

SLA surpasses other SFF techniques with respect to accuracy and resolution, achieving an accuracy of up to 20 microns [12]. The major drawback of SLA is its relatively small product size and high cost [2].

4.3 ADDITIVE MANUFACTURING MATERIALS

All engineering materials are broadly classified as metals, ceramics and polymers based on their bonding and structure. Composites combine the beneficial features of metals, ceramics, and polymers with a matrix and reinforcement constituting these material categories. The commonly used materials for AM under each category are discussed in the following sections.

4.3.1 METALS

Metallic parts are typically manufactured using conventional methods such as casting, forming and machining. Being a tool-free, cost-effective and digital approach, AM has significant advantages to offer for manufacturing metals and bring about a paradigm shift in aerospace, automobile, electrical and medical industries. Some of the key benefits of the same are [15]:

   i. The part characteristics obtained are superior to their cast and comparable to their forged counterparts.

ii. The development of complex 3D geometries such as architecture lattice constructions, topologically optimized structures and recesses for cooling channels is facilitated, which is not easy to achieve using traditional processes.

iii. Fabrication time is shortened, and the cost is reduced due to the elimination of production and assembly steps, thus cutting down material wastage and negative environmental impact.

iv. Low volume production and manufacturing of personalized products is possible.

v. 3D functionalization and surface engineering are permitted.

Metals, to be manufactured using AM, must possess two main properties: a decent ability to be welded to evade crack formation during solidification and metal stock being available in the form of spherical powders to obtain good packing density and homogeneity of the powder deposition. As of 2017, fewer than 50 different compositions of alloys are available in the form of atomized powders. Commonly used metals for AM are steel and its various types, aluminium alloys, superalloys of nickel, titanium and its alloys, precious metals (gold and silver) and refractory metals (tungsten and tantalum) [15]. These metals are pulverized to form powders using water, gas or plasma atomization before being used for the AM process. Powder characteristics, particle size, morphology and chemical composition, to name a few, vary with the difference in the powder production method. Powder characteristics influence the fabricated components' bulk material properties, such as part density and porosity [16].

Steel being the most common engineering material, is of great interest in AM. Spierings & Levy studied the AM of stainless steel 316L produced using SLS. They considered three different powder particle size distributions and varying energy densities of the laser beam and compared the resulting solid part density. They concluded that a minimum amount of fine grains is needed to fill the void between coarse grains. In the absence of those fine grains, lower scan speeds would be required to produce dense parts. Further, powder type also affected the mechanical strength, ductility and surface quality of the part [17]. Burkert & Fischer investigated the effect of various SLS process parameters with respect to the void formation in the microstructure of Marage 300, a type of maraging steel. They found energy fed following the scanning speed, thickness of the layers, and the platform's preheating to be the most influential parameters [18]. Liang *et al.* successfully repaired a 34CrNiMo6 steel sample using laser melting based AM techniques, performed heat treatment on it and studied its microstructure and mechanical properties. They observed an enhancement in the heat-treated AM sample's tensile strength, ductility, and microhardness [19]. Cui *et al.* printed a 24CrNiMo low alloy steel using laser-based AM with the aim of studying the influence of volumetric energy density on the hardness, microstructure and densification characteristics such as internal defects and surface roughness of the printed material. The microstructure largely comprised of the α ferrite phase with the presence of martensite and bainite mixture. Larger amount and size of bainite was found with a rise in laser energy density. Further, they observed an increase in microhardness of the printed alloy steel material with an increase in the energy density and concluded that internal defects primarily controlled the microhardness [20]. Napolitano *et al.* optimized the parameters for the selective laser melting of 316L stainless steel by investigating the mechanical properties, microstructure, defects and melt pool characteristics of the fabricated part. They introduced a dimensionless quantity called power ratio to define local melting intensity and found its optimum value to be around 3.5. They determined that poorer ductility and strength below the optimum resulted due to the dearth of melt fusion while the poorer properties above the optimum value are achieved owing to fusion line defects and grain growth [21]. The allotropic nature exhibited by iron alloys like steel combined with the large temperature gradient of AM opens the possibility of generating unique microstructures. On the contrary, alloys producing varied phases depending upon the cooling rate, like retained austenite and martensite in precipitation-hardened steel, impact AM parameters' choice [22].

A limited number of aluminium alloys are available for AM since aluminium is easy to machine and less expensive and does not give a high commercial advantage when processed using AM [23]. Also, aluminium alloys lack good weldability. High-performance alloys usually acquire strength as a result of precipitation hardening. In some cases, hardenable alloys may consist of elements such as zinc, which are highly volatile, which may cause turbulence in melt pools, porosity and splatter. Therefore, they are not fit for use in AM [24]. Molten aluminium's low viscosity limits AM to small-sized melt pools. However, aluminium's high thermal conductivity favours the fabrication of AM parts since it lessens stresses induced thermally, eliminates the requirement of support structures, and allows higher processing speeds [16]. Aluminium alloys commonly being used for AM include hardenable AlSi10 Mg (EN AC-43000), Al-Cu (2139) and Al-Mg-Si (6061) alloys and the eutectic AlSi12 (EN AC-44200). Schmidtke *et al.* investigated the use of aluminium alloy with scandium content for AM techniques employing lasers since the amount of scandium in modified aluminium alloys, greater than the eutectic composition, results in a material having great potential for aerospace applications. Further, alloying with magnesium or zirconium facilitates the reduction of weight and corrosion resistance and improves metallic lightweight alloys' strength properties [25]. Zhang *et al.* examined the influence that process parameters have on the properties of a sample composed of 9 % Al powder mixture using SLS. They observed a fine equiaxed grain structure and achieved higher density under lower laser input and scanning speed [26]. Yang *et al.* successfully fabricated a novel aluminium alloy i.e., the Al-5Mg$_2$Si-2Mg alloy using selective laser melting and obtained products with minimal defects and free from hot cracking. Further, they attained excellent mechanical properties including ultimate tensile strength, yield strength, elongation and a refined microstructure [27].

The most crucial feature of titanium and its alloys is its high strength to weight ratio and have wide industrial application in high-performance parts. Thus, they are of great research interest for the AM process. The large variety of titanium alloy compositions and microstructures and its allotropy combined with high-temperature gradients favours titanium alloys for AM given the complex thermal cycle involved in AM. Attar *et al.* manufactured AM parts using commercially pure titanium (CP-Ti) since it is a widely used biomedical application material. They fabricated CP-Ti parts using SLS and compared its mechanical properties with traditionally manufactured

CP-Ti parts and observed significant improvement in both the compressive and tensile ultimate strength of the sample and its microhardness, compared to the conventionally produced sample because of martensitic α' grains forming and subsequent grain refinement [28]. Carroll *et al.* investigated a cruciform shaped Ti-6Al-4V part made by laser-based AM techniques for its anisotropic mechanical properties. They achieved significant ductility in transverse and longitudinal directions. Also, oxygen being present in trace amounts was found to increase its ultimate tensile and yield strength [29]. Kolomiets *et al.* researched the application of Ti-AM in designing and manufacturing wind musical instruments using FDM. They concluded that Ti-AM is a promising technique to produce musical instruments which are cost effective, biocompatible and have unique sound properties [30].

Other metals, including nickel-based super alloys like Inconel 718, magnesium, refractory metals like tantalum and precious metals like gold have successfully been produced using AM techniques.

4.3.2 Ceramics

Ceramics demonstrate superior properties such as strength, resistance to wear, outstanding chemical inertness and high-temperature stability. Therefore, they find commercial applications in aerospace, biomedical, machine tools, electronic industries, and similar high-end applications. Ceramics are challenging to manufacture using conventional processes owing to their melting point and hardness. Hot pressing, slip casting, sintering and grinding are some ceramic manufacturing techniques. However, they result in high cost and energy consumption and grinding leads to damages caused by pulverization and micro cracking [31]. A substantial part of research in ceramic AM is focused on porous structures because the development of complex-shaped porous architectures is possible using AM alone, which gives precise control over-dimension, shape and amount of pores [32]. Some commonly used ceramics for AM include alumina (Al_2O_3), zirconia (ZrO_2), hydroxyapatite (HA), nitride aluminium and porcelain.

Wang developed a solvent-based slurry SLA and sintering technique to fabricate parts using Al_2O_3 powder and successfully built parts with a mean density of approximately 98 percent. Further, the parts built were free from delamination and cracks, hence had better mechanical properties [33]. Li *et al.* prepared Al_2O_3 ceramic, porous and mixed with CaSO4 and dextrin, using a combination of 3D printing and sintering techniques. They observed the Vickers hardness of the prepared ceramic sample to reduce with an increasing amount of $CaSO_4$ dextrin [34]. Alumina ceramics have excellent mechanical properties, are porous and has well—connected channels, along with a uniform microstructure, and can be used effectively for engineering applications. ZrO_2-Al_2O_3 alloys have enhanced toughness, superior resistance to corrosion and temperature, mechanical properties that can be controlled and excellent biocompatibility. Thus, ZrO_2-Al_2O_3 finds industrial applications in the biomedical, chemical and high-end engineering domains. Yves-Christian *et al.* studied the AM of a eutectic powder mixture of Al_2O_3-ZrO_2 using SLS. They observed that preheating the material just below its melting point ensured the elimination of cracks. The resulting solid showed a fine-grained microstructure and high flexural strength;

however, bending strength was lower than that of an equivalent conventionally manufactured part [35]. Liu *et al.* adopted laser-based 3D printing techniques to fabricate a directionally solidified eutectic Al_2O_3- ZrO_2 ceramic. They observed an irregular morphology in the microstructure of the eutectic ceramic. They obtained the fracture toughness in the transverse and longitudinal directions and found the toughness along the transverse direction to be higher. Another interesting observation was the crack length. The crack length parallel to the direction of extension of the eutectic colonies formed in the microstructure was longer than its equivalent perpendicular to the extension's direction [36]. Bertrand *et al.* manufactured 3D solid objects made of pure yttria-zirconia powders using SLS. They deduced that SLS could be successfully used to fabricate yttria-zirconia components without the addition of additives. They further studied the impact of laser length and its orientation on the quality of the manufactured product and determined that shorter laser lengths lead to droplet formation and hence should be avoided. However, the mechanical properties, especially the product's density, were much lower than the requirement for its potential, in biomedical applications, in particular [37]. Schlacher *et al.* developed a lithography-based AM technique to manufacture alumina ceramics with a characteristic strength greater than 1GPa, which is far more than the 650 MPa strength of monolithic alumina. They sandwiched alumina-zirconia layers between pure alumina ceramic layers to fabricate the part layer by layer and achieved good biaxial strength in the part [38]. Hofer *et al.* used a lithography-based 3D printing technique to print highly textured alumina ceramics. They achieved high quality texture and comparable characteristic strength to equiaxed alumina ceramics [39].

Hydroxyapatite is a calcium phosphate-based ceramic ($Ca_{10}(PO_4)_6(OH)_2$) generally used in biomedical applications like spacers, fillers and as synthetic bone graft substitutes. The primary application of HA is in dentistry and orthopaedics, primarily due to its bioactivity, biocompatibility, and osteoconduction characteristics to host tissue [40]. Scalera *et al.* built custom-made scaffolds made of different HA powder concentrations filled in epoxy UV curable suspensions for bone tissue engineering using the SLA method. They obtained good mechanical properties and low shrinkage, and the parts built showed no signs of delamination. Furthermore, they observed a fall in the reaction rate on raising the proportion of HA powder in the suspension [41]. HA components can be manufactured using SLS; however, the SLS of pure HA has not been done. HA is used in polymer matrix composites as filler and produced using SLS. Salmoria *et al.* used HA as reinforcement in high-density polyethylene (HDPE) and fabricated functionally graded HDPE/HA scaffolds using SLS. They concluded that the HA content in HDPE results in increased distance between HDPE particles and heat dissipation, affecting the degree of sintering. The flexural and ultimate strength of the built component was close to the bone tissue properties. Reinforcement of HA in HDPE also led to a reduction in cyclic fatigue and plastic deformation during their respective testing [42]. Savalani *et al.* examined the performance of Nd: YAG (Neodymium-doped Yttrium Aluminium Garnet) and CO_2 lasers for the SLS of HA-HDPE bio-ceramic polymer matrix composite. They deduced that the CO_2 laser had a wider operation range and better control over the material during its laser sintering than the Nd: YAG laser. Further, Nd: YAG laser could effectively sinter particles as small as 212 microns as opposed to 106 microns in the case of the CO_2 laser. They also determined the binding mechanism dominant when using the two types of lasers, i.e. necking in CO_2 laser and coalition of particles in Nd: YAG laser [43]. Feng *et al.* prepared HA bio-ceramic scaffolds using AM methods and studied its biocompatibility and mechanical properties. They concluded that the developed AM technique can be effectively used for the AM of HA bio-ceramics as the degradation behaviour and compression strength achieved showed potential for applications in bone tissue engineering [44]. Ceramics, such as Titanium carbide (TiC) and other titanium-based ceramics and Silicon carbide (SiC), have favourably been applied to build components using AM methods.

4.3.3 POLYMERS

Polymer AM gained attention owing to its capability to yield geometrically complex parts. Engineering polymers usually have the following properties: high strength to weight ratios (lightweight but comparatively robust), corrosion resistance, toughness, lack of electrical and thermal conductivity, resilience, colour, and low cost [45]. Hence, polymers are one of the most widely used classes of materials in AM. They find wide applications in industries, such as aerospace, architecture, toy manufacturing, and medicine. Several polymers such as acrylonitrile butadiene styrene (ABS), polylactic acid (PLA), acrylic, polycarbonate (PC), polyester, polyurethane and polystyrene are used to manufacture components using AM methods including SLS, material jetting, SLA and FDM [46]. These polymeric materials are utilized for AM as thermoplastic filaments, reactive monomers, resin or powder.

PLA is an easy to print polymer that provides good visual quality and is very rigid, strong, but brittle. It is typically fabricated using the FDM process. Song *et al.* additive manufactured PLA using the FDM method and studied its mechanical properties along different directions. They observed anisotropic and asymmetric mechanical properties and direction-dependent fracture behaviour. The mechanical response of PLA so manufactured was better than its injection moulded counterpart. Moreover, they construed that AM helps increase the crystalline nature of the material, reduce ductility, and improve fracture toughness [47]. Chacon *et al.* investigated the influence of FDM parameters, namely feed rate, build orientation, and layer thickness, on the PLA sample's mechanical properties. Based on their study, they established a set of guidelines for the FDM of PLA. They deduced that the on-edge build orientation is suitable for achieving optimal mechanical properties such as ductility, stiffness and strength. Layer thickness and the feed rate should be high to minimize printing time. Additives are also mixed into the pure PLA to help meet PLA's strength requirements; hence, PLA-based composites are commonly used in AM [48]. Hinchcliffe *et al.* considered an additively manufactured natural fibre reinforced PLA matrix composite, using plant-based fibres, for example, jute and flax and studied the impact of fibre type, matrix's cross-section, amount of reinforcement strands, and level of initial fibre prestress on the composite's flexural and tensile properties. Flax fibre surpassed jute fibres with respect to tensile properties. Also, applying a prestress force on the fibre led to the enhancement of mechanical properties [49]. Dave *et al.* printed a PLA sample using FDM to study the influence of parameters such as infill density and pattern at different sample orientations on its tensile strength. They obtained

Materials for Additive Manufacturing

greater tensile strength when samples were constructed with flat and long edge part orientation than those printed with short edge orientation during printing with concentric and rectilinear and pattern. Further, tensile strength and infill density was found to be directly proportional and maximum strength was attained at 100 percent infill density [50]. Zhou *et al.* 3D printed a carbon nanotube (CNT)/PLA composite and carried out its study using electron microscopy. They concluded that the adopted 3D printing technique helps eliminate impurities from the material. The addition of CNT in PLA aids increase crystalline nature, improves tensile strength and surface roughness [51].

ABS is a polymer with resistance to high temperature and has excellent toughness. Dana *et al.* used polymer AM techniques to print ABS components and study print direction's influence on its tensile properties. On analysis, they observed that a network of pores was formed, the structure of which depended on the printing path, and the distance between successive layers depended on the angle between them. An alteration of 90° in the printing angle between successive layers resulted in the reduction of distances and the overall fraction of pores by volume. This also led to the material possessing good tensile properties. However, printing with 0° printing angle resulted in an ABS sample with the poorest ultimate tensile strength, elongation during fracture and stiffness [52]. The surface quality of ABS products manufactured using FDM was found to be substandard and unfit for engineering applications. Hence, Kuo *et al.* devised a polishing mechanism for FDM fabricated ABS using acetone vapours to overcome the limitation of poor surface finish. The proposed technique was quick and cost-effective and reduced average surface roughness by around 98 percent. Consequently, the manufactured ABS had better dimensional accuracy and also possessed high stability [53]. Torrado *et al.* evaluated the result of mixing different additives to the ABS matrix of additively manufactured ABS samples' mechanical properties. They aimed to improve anisotropic properties and hence, created a blend of ABS with UHMWPE, i.e., ultra-high molecular weight polyethylene and SEBS, i.e., styrene ethylene butadiene styrene. The resulting combination served the purpose of improving anisotropy by creating a stronger bond and increasing the surface area in contact between consecutive layers [54]. Chawla *et al.* tested the feasibility of using a secondary recycled ABS filament as a raw material for FDM. They produced the ABS filament using extrusion and varied the extrusion parameters to study its effect on filament properties. They concluded that the use of recycled ABS filament for FDM is feasible and obtained desirable mechanical properties in the filament [55]. Nabipour and Akhoundi used FDM to print an ABS matrix composite with copper particle filler and investigated the effect of layer height, nozzle temperature, nozzle diameter and raster angle on the production time, density and tensile strength of the printed material. They determined optimum parameter settings in order to achieve maximum material properties [56].

The thermoplastic polymer, polyether ether ketone or PEEK, is a semicrystalline polymer that has outstanding mechanical, thermal and chemical properties. It is stable at high temperatures and has good chemical stability. Rinaldi *et al.* surveyed PEEK usage for space applications as a structural material, and successfully 3D printed a nanosat structure using FDM. Following a comprehensive mechanical and thermal analysis of the printed component, they concluded that PEEK has the potential to be used for space applications [57].

Polymers have the advantage of being recyclable. Boparai *et al.* synthesized a nanocomposite material using waste Nylon6-Al-Al$_2$O$_3$ using FDM. They obtained substantial improvements in the synthesized material's thermal properties as the nanofiller material in the Nylon6 matrix absorbed the crystallization heat that hindered the crystal nucleation and growth, thus acting as a thermodynamic sink. The filler also improved the thermal resistance of the material. Further, they compared the manufactured nanocomposite properties with ABS and concluded that it could be used as an alternative material for ABS [58].

Several other polymers like polycarbonate, polyethylene terephthalate have been successfully manufactured using AM methods. They also used reinforcement materials such as glass, carbon, and Kevlar fibres as polymer matrix composites.

4.4 APPLICATIONS

AM finds a wide range of utilization in the manufacturing industry today, with a plethora of materials being used to print 3D solid components for domestic and engineering applications. The automobile, aerospace, architecture, biomedical, and energy sector are few places where AM is popularly used. Some of the trending industrial applications are discussed in the following.

4.4.1 AEROSPACE

AM methods are fit for utilization in the aerospace industry, owing to its features discussed as under [3]:

   i. Complex geometry: Aerospace functions require parts with complicated shapes, components capable of performing multiple functions such as forming the structure, ensuring heat dissipation and proper flow of air, and electrical and electronic components. Hence, AM has been effectively used to manufacture fuel nozzles, fan blades and printed electronic items.

   ii. Difficult to machine materials: State-of-the-art and expensive materials such as high-strength steel alloys, super alloys of nickel, titanium and titanium alloys, ultra-high temperature ceramics find uses in this sector. AM ensures that these materials can be manufactured with ease and with minimal material wastage.

   iii. Customized production and on-demand manufacturing: Parts produced for use in the aerospace sector are made in small batches. Also, aerospace vehicles have a long working period. Hence, AM allows the production of parts in small batches, whenever required, cutting down the requirement of maintaining an inventory and the associated costs.

Turbine blades, aero-engine components, and heat exchangers have been manufactured with metallic and non-metallic materials using AM techniques such as SLA

Materials for Additive Manufacturing

and FDM. Structural components designed for the Boeing 787 Dreamliner were built with titanium using AM by a firm called 'Norsk Titanium AS'. A cost reduction of around \$2 to \$3 million per aircraft was reported by them. Another organization called 'Arconic' has fabricated fuselage and engine pylon components made of titanium for the Airbus A320 and A350 XWB test versions and spacecraft vents of NASA's SLS/Orion. A company named 'Stratasys' used FDM to produce tools and parts with polymer, ceramic and composite materials for various aerospace establishments, such as NASA, Piper Aircraft, and Bell Helicopter. Over 70 lightweight yet strong components for the Mars rover by NASA were fabricated by Stratasys using FDM. Bell Helicopter successfully slashed the manufacturing time from 6 weeks to 2.5 days by producing wiring conduits containing polycarbonate (PC) for the V-22 Osprey using FDM. The NASA Aeronautics Research Institute is also attempting to build a gas engine using lightweight, temperature resistant materials such as ceramic matrix, polyetherimide and polymer matrix composites. In addition, AM allows the repair of the high-performance components, which have been subjected to damage due to corrosion, cracking, impact or cyclic stresses [3].

Several other materials like ultra-high temperature and ultra-high strength ceramics, composites and functionally graded materials are being explored by researchers for potential application in the aerospace sector.

4.4.2 Biomaterials

Biomaterials have specific requirements which can be fulfilled efficiently by employing AM methods [3]:

i. High complexity: Medical implants, engineered organs and tissues, and controlled drug delivery mechanisms are complicated to implement and require innovative solutions. AM can aid in creating the necessary complex parts using novel engineering materials, for example, semicrystalline polymer composites.

ii. Patient-specific customization: Biomedical applications like implants, prosthetics, and drug dosage have to be designed as per the patients' needs, which vary from person to person. AM offers the scope of personalizing the needed parts and adapting to the patient's body type.

iii. Easy public access: AM builds the desired part from a CAD model, which can be easily shared among researchers. The parts can then be reproduced using the CAD model from one location to another based on the needs.

Biomaterials have been used to build parts such as bones, aortic valves, cartilage, and branched vascular trees. Keriquel *et al.* used nano-hydroxyapatite to 3D print bones using an in-vivo CAD-CAM set up and highlighted the possibility of utilizing robotics and computer assistance in the medical field [59]. Cui *et al.* 3D printed human cartilage tissue and repaired osteochondral plugs with polyethylene glycol dimethacrylate (PEGDMA) and human chondrocytes, using AM [60]. Erbel *et al.* used biodegradable magnesium to make stents which gave results similar to its metal equivalent. Magnesium stents had an additional advantage of safe degradation after about four months of use [61]. Biocompatible titanium alloys like Ti6Al4V have been used for bone tissue growth. AM was also sanctioned by the Food and Drug Administration (FDA) in 2015 for its application in drug manufacturing and delivery systems. Goyanes *et al.* used Flex EcoPLA™ (FPLA) produced commercially and polycaprolactone (PCL) filaments filled with salicylic acid to develop masks/patches with anti-acne drugs, using FDM. Better shape and characteristics were obtained with FPLA-salicylic acid as compared to PCL-salicylic acid [62]. Huang *et al.* designed drug implants consisting of levofloxacin (LVFX) and an optimized binder solution of ethanol and acetone to be used for complex drug release profiles. AM helped achieve more sophisticated drug release profiles than conventional manufacturing techniques [63]. Goyanes *et al.* used FDM to fabricate oral drug delivery devices comprising polyvinyl alcohol (PVA) filaments with paracetamol or caffeine [64].

Functional composites, including ceramic-coated metal implants, AM of artificial organs, organs being paired with electronic devices, bionic ear, for example, are being researched.

4.5 FUTURE SCOPE

The field of AM has witnessed tremendous advancements in the past few decades. Several AM techniques have been developed, and a wide range of materials are being used to additively manufactured products in various industries, including aerospace, automotive, architecture, biomedical and fashion. Further, AM is becoming increasingly popular in the manufacturing sector since it yields products with complex shapes and high precision, allows the use of versatile materials, and offers a sustainable and environmentally friendly option as it reduces energy usage and material wastage. The major limitation of AM is its high cost and production time when compared to conventional manufacturing techniques, and researchers are working to alleviate this issue [65–66].

'Industry 4.0' is a relatively recent development that aims to combine the Internet of Things (IoT) with manufacturing and make the process smarter. The amalgamation of IoT and AM in industries can optimize the AM process and increase its efficacy. The AM principles and experiences gained in research institutes must be passed to businesses through guided visits to leading facilities, guided training, induction and practical training, diagnosis, testing, prototypes and specific advice in the near future. This will enhance the fabrication of innovative and customized products and enable to achieve Demand-Driven Manufacturing, i.e., DDM, which shall further reduce energy consumption and material wastage [66].

In addition, researchers are working to make the AM process cost-effective and widen the range of materials being used and make the materials readily available [67]. Significant strides have been made in the use of composite materials for AM. Fibre-reinforced polymer composite and ceramic reinforced metal matrix composite parts are made using AM processes. The AM of composites, for example, ceramic reinforced nickel matrix composites and TiC added invar, are being researched [31].

Further, the advent of smart materials has the potential to revolutionize the field of AM. Smart materials such as shape memory alloys (SMA), shape memory polymers (SMP) and piezoelectric materials have the ability to respond to external stimuli, reconfigure and acquire necessary properties with time. SMAs have super elasticity and thermal shape recovery properties, which can be tapped in biomedical and micro-electromechanical applications. The use of SMPs in the medical and fashion industries is being researched. There are numerous prospective future smart material applications, including but not limited to lightweight configurations, programmable materials, stimulus-activated mechanisms and self-assembling structures [67]. Elahinia *et al.* additively manufactured a high temperature SMA using NiTiHf powder and investigated its functional and thermal properties. Further, they compared the NiTiHf material produced using AM and extrusion and observed the transformation stresses of the two to be comparable [68]. Cersoli *et al.* 3D printed SMP based polyurethane and analysed its mechanical properties, recovery ability and actuating performance. The printed SMP exhibited excellent flexural and ultimate tensile strengths and yielded a recovery of 96 percent and 90.7 percent shape retention. Also, they combined the developed SMP with SMA and successfully used it as a thermal switch and actuated an electronic circuit [69].

The introduction of smart materials has also led to the creation of another class of AM called 4D printing. The fourth dimension in 4D printing includes time, shape, functionality or how a property of a 3D printed component could change with time when given an external stimulus in the form of heat, light, water or pH [70]. Four-dimensional printing allows programming at the material level. Materials are directly embedded with sensing and actuation. Self-assembly is done at a macro scale to eliminate post-manufacturing assembly. This mode of printing also reduces the number of parts and volume of shipping, and these materials can be used for bioinspired morphing features [71]. Materials for 4D printing are being developed to be used as a coating for defence vehicles such that they can adapt with changes in the environment, and 4D printing of antennae and solar cells for satellites and space vehicles is being explored. Biomaterials used in 4D printed have a huge potential for application in the medical field in bone tissue engineering and self-assembling parts. Components of robots such as motors and sensors can be replaced by 4D materials for reduction of cost and size [72]. Thus, 4D printing of materials is a beneficial process and is being researched for its industrial applications.

4.6 CONCLUSION

This chapter attempts to recapitulate some of the existing AM techniques and highlights some of the prominent materials used with these techniques, which fall under three broad categories: metals, ceramics and polymers. AM is gaining popularity in the manufacturing sector, and researchers have developed several AM processes and materials for this sector over the years. Given the numerous advantages of AM, it has commercial applications in the medical, aerospace, automotive industries. The near future shall witness significant innovations in AM, with AM becoming a mainstream manufacturing process that replaces traditional methods.

List of Acronyms

Acronym	Full Form
ABS	acrylonitrile butadiene styrene
AM	additive manufacturing
ASTM	American Society for Testing Materials
CAD	computer aided design
CAM	computer aided manufacturing
CC	contour crafting
CNT	carbon nanotubes
CP-Ti	commercially pure titanium
DDM	demand driven manufacturing
FDM	fused deposition modelling
FPLA	Flex EcoPLA
HA	hydroxyapatite
HDPE	high density polyethylene
IoT	Internet of Things
LOM	laminated object manufacture
LVFX	levofloxacin
Nd:YAG	neodymium-doped yttrium aluminium garnet
PC	polycarbonate
PCL	polycaprolactone
PEEK	polyether ether ketone
PEGDMA	polyethylene glycol dimethacrylate
PLA	polylactic acid
PVA	polyvinyl alcohol
SEBS	styrene ethylene butadiene styrene
SFF	solid free form
SGC	solid ground curing
SLA	stereolithography
SLS	selective laser sintering
SMA	shape memory alloy
SMP	shape memory polymer
STL	standard tessellation language
Ti-AM	titanium additive manufacturing
UHMWPE	ultra-high molecular weight polyethylene

REFERENCES

[1] J. J. Beaman, D. L. Bourell, C. C. Seepersad, and D. Kovar, "Additive manufacturing review: Early past to current practice," *J. Manuf. Sci. Eng.*, vol. 142, no. 11, pp. 1–20, 2020, doi: 10.1115/1.4048193.

[2] S. H. Huang, P. Liu, A. Mokasdar, and L. Hou, "Additive manufacturing and its societal impact: A literature review," *Int. J. Adv. Manuf. Technol.*, vol. 67, no. 5–8, pp. 1191–1203, 2013, doi: 10.1007/s00170-012-4558-5.

[3] T. D. Ngo, A. Kashani, G. Imbalzano, K. T. Q. Nguyen, and D. Hui, "Additive manufacturing (3D printing): A review of materials, methods, applications and challenges," *Compos. Part B Eng.*, vol. 143, December 2017, pp. 172–196, 2018, doi: 10.1016/j.compositesb.2018.02.012.

Materials for Additive Manufacturing

[4] D. Bourell *et al.*, "Materials for additive manufacturing," *CIRP Ann.—Manuf. Technol.*, vol. 66, no. 2, pp. 659–681, 2017, doi: 10.1016/j.cirp.2017.05.009.

[5] I. Gibson, D. W. Rosen, and B. Stucker, *Additive Manufacturing Technologies Rapid Prototyping to Direct Digital Manufacturing.* Springer, 2010.

[6] R. Singh and H. K. Garg, *Fused Deposition Modeling—A State of Art Review and Future Applications.* Elsevier Ltd., 2016.

[7] P. Kumar and I. P. S. Ahuja, "Application of fusion deposition modelling for rapid investment casting—a review," *Int. J. Mater. Engg. Innov.*, vol. 3, pp. 204–227, 2012.

[8] A. Mazzoli, "Selective laser sintering in biomedical engineering," *Med. Biol. Eng. Comput.*, vol. 51, no. 3, pp. 245–256, 2013, doi: 10.1007/s11517-012-1001-x.

[9] S. Singh, V. S. Sharma, and A. Sachdeva, "Progress in selective laser sintering using metallic powders: A review," *Mater. Sci. Technol. (United Kingdom)*, vol. 32, no. 8, pp. 760–772, 2016, doi: 10.1179/1743284715Y.0000000136.

[10] E. O. Olakanmi, R. F. Cochrane, and K. W. Dalgarno, "A review on selective laser sintering/melting (SLS/SLM) of aluminium alloy powders: Processing, microstructure, and properties," *Prog. Mater. Sci.*, vol. 74, pp. 401–477, 2015, doi: 10.1016/j.pmatsci.2015.03.002.

[11] A. N. Chen *et al.*, "High-performance ceramic parts with complex shape prepared by selective laser sintering: A review," *Adv. Appl. Ceram.*, vol. 117, no. 2, pp. 100–117, 2018, doi: 10.1080/17436753.2017.1379586.

[12] F. P. W. Melchels, J. Feijen, and D. W. Grijpma, "A review on stereolithography and its applications in biomedical engineering," *Biomaterials*, vol. 31, no. 24, pp. 6121–6130, 2010, doi: 10.1016/j.biomaterials.2010.04.050.

[13] P. J. Bártolo, *Stereolithography*, vol. 53, no. 9. Springer, 2011.

[14] J. Huang, Q. Qin, and J. Wang, "A review of stereolithography: Processes and systems," *Processes*, vol. 8, no. 9, 2020, doi: 10.3390/PR8091138.

[15] S. Gorsse, C. Hutchinson, M. Gouné, and R. Banerjee, "Additive manufacturing of metals: A brief review of the characteristic microstructures and properties of steels, Ti-6Al-4V and high-entropy alloys," *Sci. Technol. Adv. Mater.*, vol. 18, no. 1, pp. 584–610, 2017, doi: 10.1080/14686996.2017.1361305.

[16] D. Herzog, V. Seyda, E. Wycisk, and C. Emmelmann, "Additive manufacturing of metals," *Acta Mater.*, vol. 117, pp. 371–392, 2016, doi: 10.1016/j.actamat.2016.07.019.

[17] G. Spierings, A.B. & Levy, "Reviewed, accepted 9/15/09," *Rev. Lit. Arts Am.*, pp. 342–353, 2009.

[18] T. Burkert and A. Fischer, "The effects of heat balance on the void formation within marage 300 processed by selective laser melting," pp. 745–757, 2015. Available: http://library1.nida.ac.th/termpaper6/sd/2554/19755.pdf.

[19] R. Liang *et al.*, "Microstructure and mechanical properties of 34CrNiMo6 steel repaired by laser remelting," *J. Mater. Res. Technol.*, vol. 9, no. 6, pp. 13870–13878, November 2020, doi: 10.1016/j.jmrt.2020.09.100.

[20] X. Cui, S. Zhang, C. H. Zhang, J. B. Zhang, and S. Y. Dong, "Additive manufacturing of 24CrNiMo low alloy steel by selective laser melting: Influence of volumetric energy density on densification microstructure and hardness," *Mater. Sci. Eng. A.*, vol. 809, pp. 140957, 2021.

[21] R. E. Napolitano, S. Jain, C. Sobczak, B. A. Augustine, and E. Johnson, "Build optimization for selective laser melting of 316L stainless steel and parameterization for cross material comparison and process design," *J. Mater. Eng. Perform.*, vol. 30, no. 12, 2021, doi: 10.1007/s11665-021-05861-7.

[22] L. E. Murr *et al.*, "Microstructures and properties of 17–4 PH stainless steel fabricated by selective laser melting," *J. Mater. Res. Technol.*, vol. 1, no. 3, pp. 167–177, 2012, doi: 10.1016/S2238-7854(12)70029-7.

[23] C. Brice, R. Shenoy, M. Kral, and K. Buchannan, "Precipitation behavior of aluminum alloy 2139 fabricated using additive manufacturing," *Mater. Sci. Eng. A.*, vol. 648, pp. 9–14, 2015, doi: 10.1016/j.msea.2015.08.088.

[24] K. Bartkowiak, S. Ullrich, T. Frick, and M. Schmidt, "New developments of laser processing aluminium alloys via additive manufacturing technique," *Phys. Procedia*, vol. 12, no. PART 1, pp. 393–401, 2011, doi: 10.1016/j.phpro.2011.03.050.

[25] K. Schmidtke, F. Palm, A. Hawkins, and C. Emmelmann, "Process and mechanical properties: Applicability of a scandium modified Al-alloy for laser additive manufacturing," *Phys. Procedia*, vol. 12, no. PART 1, pp. 369–374, 2011, doi: 10.1016/j.phpro.2011.03.047.

[26] B. Zhang, H. Liao, and C. Coddet, "Effects of processing parameters on properties of selective laser melting Mg—9 % Al powder mixture," *Mater. Des.*, vol. 34, pp. 753–758, 2012, doi: 10.1016/j.matdes.2011.06.061.

[27] H. Yang, Y. Zhang, J. Wang, Z. Liu, C. Liu, and S. Ji, "Additive manufacturing of a high strength Al-5Mg2Si-2Mg alloy: Microstructure and mechanical properties," *J. Mater. Sci. Technol.*, vol. 91, pp. 215–223, 2021.

[28] H. Attar, M. Calin, L. C. Zhang, S. Scudino, and J. Eckert, "Manufacture by selective laser melting and mechanical behavior of commercially pure titanium," *Mater. Sci. Eng. A.*, vol. 593, pp. 170–177, 2014, doi: 10.1016/j.msea.2013.11.038.

[29] B. E. Carroll, T. A. Palmer, and A. M. Beese, "Anisotropic tensile behavior of Ti-6Al-4V components fabricated with directed energy deposition additive manufacturing," *Acta Mater.*, vol. 87, pp. 309–320, 2015, doi: 10.1016/j.actamat.2014.12.054.

[30] A. Kolomiets, Y. J. Grobman, V. V. Popov, E. Strokin, G. Senchikhin, and E. Tarazi, "The titanium 3D-printed flute: New prospects of additive manufacturing for musical wind instruments design," *J. New Music Res.*, vol. 50, no. 1, pp. 1–17, 2021, doi: 10.1080/09298215.2020.1824240.

[31] Y. Hu and W. Cong, "A review on laser deposition-additive manufacturing of ceramics and ceramic reinforced metal matrix composites," *Ceram. Int.*, vol. 44, no. 17, pp. 20599–20612, 2018, doi: 10.1016/j.ceramint.2018.08.083.

[32] A. Zocca, P. Colombo, C. M. Gomes, and J. Günster, "Additive manufacturing of ceramics: Issues, potentialities, and opportunities," *J. Am. Ceram. Soc.*, vol. 98, no. 7, pp. 1983–2001, 2015, doi: 10.1111/jace.13700.

[33] J. C. Wang, "A novel fabrication method of high strength alumina ceramic parts based on solvent-based slurry stereolithography and sintering," *Int. J. Precis. Eng. Manuf.*, vol. 14, no. 3, pp. 485–491, 2013, doi: 10.1007/s12541-013-0065-3.

[34] Y. Li, Y. Hu, W. Cong, L. Zhi, and Z. Guo, "Additive manufacturing of alumina using laser engineered net shaping: Effects of deposition variables," *Ceram. Int.*, vol. 43, no. 10, pp. 7768–7775, 2017, doi: 10.1016/j.ceramint.2017.03.085.

[35] H. Yves-Christian, W. Jan, M. Wilhelm, W. Konrad, and P. Reinhart, "Net shaped high performance oxide ceramic parts by Selective Laser Melting," *Phys. Procedia*, vol. 5, no. PART 2, pp. 587–594, 2010, doi: 10.1016/j.phpro.2010.08.086.

[36] Z. Liu *et al.*, "Microstructure and mechanical properties of Al2O3/ZrO2 directionally solidified eutectic ceramic prepared by laser 3D printing," *J. Mater. Sci. Technol.*, vol. 32, no. 4, pp. 320–325, April. 2016, doi: 10.1016/j.jmst.2015.11.017.

[37] P. Bertrand, F. Bayle, C. Combe, P. Goeuriot, and I. Smurov, "Ceramic components manufacturing by selective laser sintering," *Appl. Surf. Sci.*, vol. 254, no. 4, pp. 989–992, 2007, doi: 10.1016/j.apsusc.2007.08.085.

[38] J. Schlacher *et al.*, "Additive manufacturing of high-strength alumina through a multi-material approach," *Open Ceram.*, vol. 5, p. 100082, 2021, doi: 10.1016/j.oceram.2021.100082.

[39] A. K. Hofer, I. Kraleva, and R. Bermejo, "Additive manufacturing of highly textured alumina ceramics," *Open Ceram.*, vol. 5, 2021, doi: 10.1016/j.oceram.2021.100085.

[40] L. Ferrage, G. Bertrand, P. Lenormand, D. Grossin, and B. Ben-Nissan, "A review of the additive manufacturing (3DP) of bioceramics: Alumina, zirconia (PSZ) and hydroxyapatite," *J. Aust. Ceram. Soc.*, vol. 53, no. 1, pp. 11–20, 2017, doi: 10.1007/s41779-016-0003-9.

[41] F. Scalera, C. Esposito Corcione, F. Montagna, A. Sannino, and A. Maffezzoli, "Development and characterization of UV curable epoxy/hydroxyapatite suspensions for stereolithography applied to bone tissue engineering," *Ceram. Int.*, vol. 40, no. 10, pp. 15455–15462, 2014, doi: 10.1016/j.ceramint.2014.06.117.

[42] G. V. Salmoria, E. A. Fancello, C. R. M. Roesler, and F. Dabbas, "Functional graded scaffold of HDPE/HA prepared by selective laser sintering: Microstructure and mechanical properties," *Int. J. Adv. Manuf. Technol.*, vol. 65, no. 9–12, pp. 1529–1534, 2013, doi: 10.1007/s00170-012-4277-y.

[43] M. Savalani, L. Hao, and R. A. Harris, "Evaluation of CO2 and Nd:YAG lasers for the selective laser sintering of HAPEX," *Proc. Inst. Mech. Eng. Part B J. Eng. Manuf.*, vol. 220, no. 2, pp. 171–182, 2006, doi: 10.1243/095440505X32986.

[44] C. Feng *et al.*, "Additive manufacturing of hydroxyapatite bioceramic scaffolds: Dispersion, digital light processing, sintering, mechanical properties, and biocompatibility," *J. Adv. Ceram.*, vol. 9, no. 3, pp. 360–373, 2020, doi: 10.1007/s40145-020-0375-8.

[45] L. C. Brinson and H. F. Brinson, *Polymer Engineering Science and Viscoelasticity: An Introduction—Characteristics, Applications and Properties of Polymers*, Springer: New York, pp. 56–97, 2015.

[46] I. Jasiuk, D. W. Abueidda, C. Kozuch, S. Pang, F. Y. Su, and J. McKittrick, "An overview on additive manufacturing of polymers," *Jom*, vol. 70, no. 3, pp. 275–283, 2018, doi: 10.1007/s11837-017-2730-y.

[47] Y. Song, Y. Li, W. Song, K. Yee, K. Y. Lee, and V. L. Tagarielli, "Measurements of the mechanical response of unidirectional 3D-printed PLA," *Mater. Des.*, vol. 123, pp. 154–164, 2017, doi: 10.1016/j.matdes.2017.03.051.

[48] J. M. Chacón, M. A. Caminero, E. García-Plaza, and P. J. Núñez, "Additive manufacturing of PLA structures using fused deposition modelling: Effect of process parameters on mechanical properties and their optimal selection," *Mater. Des.*, vol. 124, pp. 143–157, 2017, doi: 10.1016/j.matdes.2017.03.065.

[49] S. A. Hinchcliffe, K. M. Hess, and W. V. Srubar, "Experimental and theoretical investigation of prestressed natural fiber-reinforced polylactic acid (PLA) composite materials," *Compos. Part B Eng.*, vol. 95, pp. 346–354, 2016, doi: 10.1016/j.compositesb.2016.03.089.

[50] H. K. Dave, N. H. Patadiya, A. R. Prajapati, and S. R. Rajpurohit, "Effect of infill pattern and infill density at varying part orientation on tensile properties of fused deposition modeling-printed poly-lactic acid part," *Proc. Inst. Mech. Eng. Part C J. Mech. Eng. Sci.*, vol. 235, no. 10, pp. 1811–1827, 2021, doi: 10.1177/0954406219856383.

[51] X. Zhou *et al.*, "Additive manufacturing of CNTs/PLA composites and the correlation between microstructure and functional properties," *J. Mater. Sci. Technol.*, vol. 60, 2021, doi: 10.1016/j.jmst.2020.04.038.

[52] H. Ramezani Dana, F. Barbe, L. Delbreilh, M. Ben Azzouna, A. Guillet, and T. Breteau, "Polymer additive manufacturing of ABS structure: Influence of printing direction on mechanical properties," *J. Manuf. Process.*, vol. 44, pp. 288–298, 2019, doi: 10.1016/j.jmapro.2019.06.015.

[53] C. C. Kuo, C. M. Chen, and S. X. Chang, "Polishing mechanism for ABS parts fabricated by additive manufacturing," *Int. J. Adv. Manuf. Technol.*, vol. 91, no. 5–8, pp. 1473–1479, 2017, doi: 10.1007/s00170-016-9845-0.

[54] A. R. Torrado, C. M. Shemelya, J. D. English, Y. Lin, R. B. Wicker, and D. A. Roberson, "Characterizing the effect of additives to ABS on the mechanical property anisotropy of specimens fabricated by material extrusion 3D printing," *Addit. Manuf.*, vol. 6, pp. 16–29, 2015, doi: 10.1016/j.addma.2015.02.001.

[55] K. Chawla, R. Singh, and J. Singh, "On recyclability of thermoplastic ABS polymer as fused filament for FDM technique of additive manufacturing," *World J. Eng.*, February, 2021, doi: 10.1108/WJE-11-2020-0580.

[56] M. Nabipour and B. Akhoundi, "An experimental study of FDM parameters effects on tensile strength, density, and production time of ABS/Cu composites," *J. Elastomers Plast.*, vol. 53, no. 2, pp. 146–164, 2021, doi: 10.1177/0095244320916838.

[57] M. Rinaldi, F. Cecchini, L. Pigliaru, T. Ghidini, F. Lumaca, and F. Nanni, "Additive manufacturing of polyether ether ketone (Peek) for space applications: A nanosat polymeric structure," *Polymers (Basel).*, vol. 23, no. 1, pp. 1–16, 2021, doi: 10.3390/polym13010011.

[58] K. S. Boparai, R. Singh, F. Fabbrocino, and F. Fraternali, "Thermal characterization of recycled polymer for additive manufacturing applications," *Compos. Part B Eng.*, vol. 106, pp. 42–47, 2016, doi: 10.1016/j.compositesb.2016.09.009.

[59] V. Keriquel *et al.*, "In vivo bioprinting for computer- and robotic-assisted medical intervention: Preliminary study in mice," *Biofabrication*, vol. 2, no. 1, 2010, doi: 10.1088/1758-5082/2/1/014101.

[60] X. Cui, K. Breitenkamp, M. G. Finn, M. Lotz, and D. D. D'Lima, "Direct human cartilage repair using three-dimensional bioprinting technology," *Tissue Eng.—Part A*, vol. 18, no. 11–12, pp. 1304–1312, 2012, doi: 10.1089/ten.tea.2011.0543.

[61] R. Erbel *et al.*, "Temporary scaffolding of coronary arteries with bioabsorbable magnesium stents: A prospective, non-randomised multicentre trial," *Lancet*, vol. 369, no. 9576, pp. 1869–1875, 2007, doi: 10.1016/S0140-6736(07)60853-8.

[62] A. Goyanes, U. Det-Amornrat, J. Wang, A. W. Basit, and S. Gaisford, "3D scanning and 3D printing as innovative technologies for fabricating personalized topical drug delivery systems," *J. Control. Release*, vol. 234, pp. 41–48, 2016, doi: 10.1016/j.jconrel.2016.05.034.

[63] W. Huang, Q. Zheng, W. Sun, H. Xu, and X. Yang, "Levofloxacin implants with predefined microstructure fabricated by three-dimensional printing technique," *Int. J. Pharm.*, vol. 339, no. 1–2, pp. 33–38, 2007, doi: 10.1016/j.ijpharm.2007.02.021.

[64] A. Goyanes *et al.*, "3D printing of medicines: Engineering novel oral devices with unique design and drug release characteristics," *Mol. Pharm.*, vol. 12, no. 11, pp. 4077–4084, 2015, doi: 10.1021/acs.molpharmaceut.5b00510.

[65] J. Bhattacharjya, S. Tripathi, A. Taylor, M. Taylor, and D. Walters, "Additive manufacturing: Current status and future prospects," *IFIP Adv. Inf. Commun. Technol.*, vol. 434, pp. 365–372, 2014, doi: 10.1007/978-3-662-44745-1_36.

[66] M. Jiménez, L. Romero, I. A. Domínguez, M. D. M. Espinosa, and M. Domínguez, "Additive manufacturing technologies: An overview about 3D printing methods and future prospects," *Complexity*, vol. 2019, 2019, doi: 10.1155/2019/9656938.

[67] A. Yadav *et al.*, "Investigation on the materials used in additive manufacturing: A study," 2020, https://doi.org/10.1016/j.matpr.2020.10.975.

[68] M. Elahinia *et al.*, "Additive manufacturing of NiTiHf high temperature shape memory alloy," *Scripta Materialia*, vol. 145. pp. 90–94, 2018, doi: 10.1016/j.scriptamat.2017.10.016.

[69] T. Cersoli, A. Cresanto, C. Herberger, E. Macdonald, and P. Cortes, "3D Printed shape memory polymers produced via direct pellet extrusion," *Micromachines*, vol. 12, no. 1. pp. 1–12, 2021, doi: 10.3390/mi12010087.

[70] X. Kuang *et al.*, "Advances in 4D printing: Materials and applications," *Adv. Funct. Mater.*, vol. 29, no. 2, 2019, doi: 10.1002/adfm.201805290.

[71] F. Momeni and J. Ni, "Laws of 4D printing," *Engineering*, vol. 6, no. 9, pp. 1035–1055, 2020, doi: 10.1016/j.eng.2020.01.015.

[72] A. Ahmed, S. Arya, V. Gupta, H. Furukawa, and A. Khosla, "4D printing: Fundamentals, materials, applications and challenges," *Polymer (Guildf)*, vol. 228, pp. 123926, 2021.

5 Green 3D Printing
Advancement to Sustainable Manufacturing

Amber Batwara, Harsh Mundra, Apoorva Daga, Vikram Sharma, and Mohit Makkar

5.1 Introduction 99
    5.1.1 History 100
    5.1.2 3D Printing Responds to the COVID-19 Pandemic (2020) 102
    5.1.3 Green 3D Printing 103
        5.1.3.1 Advances in Environmental Protection 105
        5.1.3.2 A Framework of Criteria and Sub-Criteria for Green 3D Printing 106
    5.1.4 Applying AHP Method to Our Framework 109
        5.1.4.1 Advantages of AHP Method 110
        5.1.4.2 Steps for AHP Method 110
    5.1.5 Data analysis and results 111
    5.1.6 Conclusion 111
    5.1.7 Limitations and Future Scope 114
References 114

5.1 INTRODUCTION

Three-dimensional (3D) printing is the process of building natural solid objects from a digital file. This process is also termed *additive manufacturing* as well as *rapid prototyping*. The terms are meant to distinguish it from subtractive processes. This additive manufacturing technology can provide the industry with increased design freedom, lower energy consumption, and shorten the time to market. Quick tooling, rapid prototyping with the prototype, direct part manufacture, and component repair of plastic, metal, ceramic, and composite materials are the most common additive manufacturing applications. The input raw material type and the energy source used to construct the product from the 3D design are the two most essential aspects of the metal additive manufacturing process.

In the last developing years, 3D printing technology continues to develop, majorly in industrial applications. Nowadays, hundreds of ready to raw materials will be used for 3D printing, and they also include some environmentally green and biodegradable materials. These materials include edible materials, ceramics, plaster, and nylon. The number of fabric types can promote the appliance of this new technology in the

DOI: 10.1201/9781003220237-5

different productive sectors. Nowadays, this 3D printing technology can produce bio-constructs, spare parts, electronics, only parts, and even jewelry [1].

How does 3D printing help in traditional production mode and industrial layout?

Because 3D printing is used in every sector nowadays, this technology helps to create a flexible and understandable layout for any machining company. Like if any company wants to adopt this technology, it is possible because due to these advancements, this technology can work with hundreds of materials, is easy to install and work fully automatic. This helps to create an excellent choice as a production model. There are multiple sectors in 3D printing which are developed to a very advanced level now, such as:

Materials Science

- Technological advancements in UV-resistant materials for stereolithography in 3D printing [2].
- Innovation of Fine graded powdered metals, which are now being used in 3D printing.
- Thermo Polymers are blended for filament deposition modeling in 3D printing.
- Innovation of materials that decomposes to a fluid and fine powder through thermal treatment is also child safe and food-safe material.
- Materials that are produced by two thermal polymerizations and are being used for highly high-resolution Printing.

Electro-Mechanical Engineering

- They are mechanically moving print heads with high acceleration and speed.
- Elimination of Torsion, Strains, and ringing of arms in machine.
- Closed-loop endpoint inspection and detection with calibration with appropriate software used.
- Ultra fine granularity of rotation and translation during machine operation.
- High strength and light-weight machine parts which are fully capable of moving with high speed without jerks

Software

- Cheaper and Better slicer software is available now in the market, and also multi-model batch slicer software is available now in the market.
- Improved algorithms for strength to weight goal-oriented humidity and temperature patterns.
- Batch software was developed to support optimization in 3D printing.

5.1.1 HISTORY

According to English philosopher Francis Bacon, learning and knowledge transformed the status of the world and marked the beginning of a new era. The Gutenberg press could produce a unique 3500 printed pages per day. This constituted the major production row, which resulted in mass production. Throughout the mid-nineties, the sector began to view signs of diversification with two different areas of impact that are more elaborated today:

- The 3D printing was still costly, which was accelerated by the growth of production. This is still ongoing—and growing.
- The outcomes are becoming perceptible in applications of productions in the aerospace, automotive, medical, and jewelry sectors as research and development and qualifications are now giving off. A significant share remains under non-disclosure agreements (NDA).

Concept modelers were developed and advanced by some of the 3D printing system manufacturers. However, these systems were all still considerably for applications of the industry.

In the early 1980s, 3D printing was originated by Charles Hull, who described stereolithography (STL) or layers of cloth to create 3D objects. Hideo Kodama of the Nagoya Municipal Industrial Research Institute (Nagoya, Japan) investigated and published the production of a printed solid model for the first time in 1981, marking the beginning of "additive manufacturing," "rapid prototyping," or "3D printing technology." Additive manufacturing" creates materials and objects, from a digital model, sequentially using an additive layering process.

The technology has progressed since Hull of 3D Systems Corp designed and realized the first 3D printer in 1984. These machines have become more valuable as their prices have dropped, making them more accessible.

Sales of 3D printing machines have increased dramatically, and their cost has steadily decreased since the turn of the century [3].

In the following years, this technology has changed into a useful gizmo. Wire frameworks in composite ear scaffolds improved the fibrous collagen matrix's mechanical properties and ensured a 3D shape after implantation. The 3D-printed ears can collect inductive signals from electrodes that describe the joining abilities of biological functions. But as scaffolds or soft tissues are supported, tissue engineering is made to accumulate actual tissues.

The outcomes reveal that it had advantages over other methods in every term. By the first decade of the 2010s, the phrases 3D printing and additive manufacturing had acquired new meanings. There were several different terms for AM technologies, which became popular among consumer-makers and the media. As a result, industrial AM end-use component manufacturers, AM machine manufacturers, and global technical standards bodies use the other. Both terminologies show the straightforward incontrovertible truth of sharing themes between technologies

In recent times, the preceding decade was when metal end-use parts like engine brackets and huge nuts were developed (rather than machining) in production instead of a plate or bar stoke machining.

Rapid prototyping features an extensive selection of applications in many fields of activities for humans. The microarchitecture reflects the tissue architecture. 3D printing is also being examined as a relevant source to replace organs; the Sales of 3D printing machines have increased dramatically, and their cost has steadily decreased since the turn of the century. Those are not functioning by unreal printed organs, such as kidneys, hearts, or skin. The treatment of disease has been advanced and removed the shortage of organ transplants [4].

It's also feasible that, in the future, drug compound databases will take the position of pharmaceutical businesses, with databases being handed to pharmacies so that they can print only the amounts of pharmaceuticals that are required. Vaccines could then be delivered through email to the pharmacy at the point of care, where they would be printed and administered.

5.1.2 3D Printing Responds to the COVID-19 Pandemic (2020)

In its early stages, 3D printing was mainly used for prototyping and tiny products development. Now, 3D printing machines are introduced to industrial applications too. New software and proper machined and mechanical kits have made 3D printing technology affordable and easy to use, and spread within the market. Today, there are various kinds of giant machines and industries that are adopting this technology because of its potential to deal with multiple technical challenges

The fourth technological revolution is creating a chance for this technology in the supply of products or services. 3D printing technologies can combine with internet-based communication systems like IoT and cyber-physical systems, which can cause the normalization of producing actions and various operations and significant supply chain modification. But revolution also brings new challenges.

The manufacturing industries for the entire production and labor pool are going to be decreased rapidly. Additionally, this technology made the market very dynamic and digital. The consumption of produce questions how different companies have to organize their operations to lead by an improved response to the extraordinary market supply and needs challenges.

NSWC Corona has been assigned as the Navy METCAL program's scientific and technical (S&T) advisor by the Chief of Naval Operations. The objective of NSWC Corona is to identify the Navy's calibration requirements and to guarantee that measuring capabilities and calibration standards are appropriately designed, implemented, and supported. Ammunition, guided weapons, and electric equipment could all be printed by Navy AM [5].

Throughout recent times, this technology of 3D printing has had a significant effect on industries and the growing world. In the year 2020, many engineers and manufacturers stepped into the fight against the COVID-19 pandemic. Therefore, they decided to use the latest technologies, including 3D printing in the community, to produce the necessary products and materials needed for medications and treatment of the current pandemic. Many Companies with 3D printing technologies pay for 3D print masks, respirators, valves, and many other parts [6]. In Figure 5.1, we can see that all the valves used with oxygen cylinders for COVID-19 treatment are being manufactured with the help of 3D printing technology only. Apart from this, 3D printing technology helps in the manufacturing of many other hospital treatment apparatus and personal protective equipment (PPE) kits, facemasks, and disinfection kits.

Green 3D Printing

FIGURE 5.1 Breathing circulation valves used in oxygen cylinders for treatment of COVID-19 are printed through 3D printing technology.

Here are some latest work lists is done by researchers in this sector

TABLE 5.1
3D Printing Applications in Various Fields

3D Printing Application	Source
Medical applications: Eyeglasses, personalized prosthetic devices, dental implants, 3D printing of organs and tissue structures.	[7, 8]
Fabric and Fashion Industry: Fiber reinforcement.	[4]
Healthcare: 3D printed skin, a print drug for increase efficiency and control of dose.	[9–12]
Smart materials: Ceramic, biomaterials, electronic materials, and composites of innovative material.	[13]
Aerospace sector: Lightweight parts, improved and complicated geometries.	[14–16]
Automotive industry: Local motors, 3D-printed bus (OLLI), and lightweight wing structures for unmanned aerial vehicles.	[17, 18]
Food industry: 3D-food printing.	[19, 20]
Architecture: Sustainable modern architecture and innovative construction.	[21, 22]
Water treatment: Water treatment technologies.	[23]
3D printing glass: Glass ornaments, optical elements, microfluidics.	[24]
Quality behaviors: 3D printed cereal-based morphologies.	[25]

5.1.3 Green 3D Printing

The problem has arisen for society as the quantity of used items is fast rising due to their qualities, impacting human health and living surroundings. Recycling, reusing, and reducing are undoubtedly the best ways to dispose of waste. Products and plastics need to be reusable and recyclable [3] and consume less energy to reduce their environmental impact [26].

Over the last few decades, 3D printing technology has evolved into a burgeoning industry and a lucrative opportunity, particularly for engineers and scientists. The additive manufacturing process instead of subtractive processes has provided a significant advantage by lessening materials and other resources. It starts from a dream of creating objects in nearby industrial facilities, and making simply the particular item that is wanted, instead of making a whole group in a removed area and delivering and warehousing the things in mass amounts. Numerous scientists think the ability to make such moulded parts and increase energy proficiency may offer the best ecological advantages from added substance production.

Since AM has been a better way to save resources even after growing the manufacturing and processes, it is called green 3D printing. Also, it has been very convenient for people to manufacture through digital designs.

Two significant advantages of 3D printing are as follows:

- Manufacturing by 3D printing produces less wastage than traditional techniques like casting, forging, cutting, and stamping.
- 3D printing in homes, factories, and the community reduces the transportation needed to get the products to final users as digital designs can be manufactured anywhere with the printers.

However, the quantitative and qualitative analysis of the environmental and green performance of 3D printing has been limited. This study mainly focuses on the energy used and the raw materials instead of focusing on the impacts of raw materials and processes and the use of waste management.

Environmental reflections are related to energy consumption and resources used, and emissions and waste. According to previous research, the effects of 3D printing are a life cycle balance with resource use having a major influence.

However, the science is uncertain about whether 3D printing has decreased or increased the environmental effects compared to other manufacturing methods. Additive manufacturing is not an intrinsically waste-less cycle. But can be improved with some changes in the process, materials, and other factors in 3D printing, which influence waste, physical features, and thus the environment.

We found that moderation in process and materials is required with a good understanding of the technology of implementation and the current state of development for improving the environment and the health of humans. Future research has stated some factors and sub-factors related to the 3D printing process, materials, inventories, and usages that affect the environment and humans. We've described their impact and how it affects green 3D printing and what is most affected, along with the framework of factors and subfactors. We've also considered which factors have the most significant impact and which have the most negligible impact, which elements need to be improved the most, and what procedures need to be modified to polish our surroundings.

Green 3D Printing

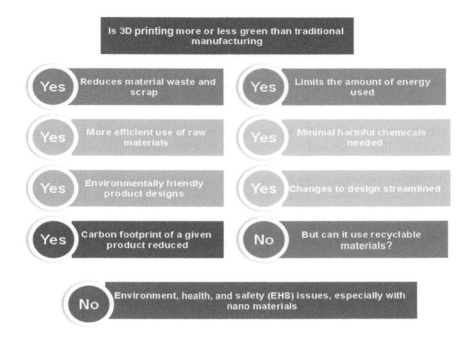

FIGURE 5.2 3D printing is more or less green than traditional manufacturing.

The performance of the electronic system improves with multi-material printing technology in Industry 4.0, allowing more unique designs to be made in only one procedure [27]. Creating a green electronic device with fewer production costs, high reliability, more safety, and rapid production is urgently needed to solve environmental problems in today's society [1]. UV-induced chain-growth and step-growth polymerization are employed in a two-step 3D-printing approach to sustainable manufacturing [2].

5.1.3.1 Advances in Environmental Protection

3D printing is achieving high performance with concerns of environmental protection and minimization of resources. Many items' transportation and production carbon footprints may be minimized if designs are sent across the globe instead of objects. These designs will be distributed digitally to individuals or businesses who will "print" the product. Furthermore, the final product's carbon footprint would significantly decrease by reducing or eliminating complex supply chains of parts made by dozens, if not hundreds, of suppliers spread across the globe. Furthermore, the final product's carbon footprint would significantly decrease by reducing or removing complex supply chains of parts made by dozens, if not hundreds, of suppliers spread across the world. It could help bridge the gap between supply and demand for nonrenewable resources, which is widening. The printing technique produces nearly no waste compared to "traditional manufacturing" and other current processes. The exact amount of raw materials will result in more finished goods, saving valuable resources.

Furthermore, 3D printing may improve recycled materials like plastics and metals, particularly for lower-end products. Excess or unsold manufacturing and the expense of inventory and spare parts storage are other sources of waste that could be drastically reduced or eliminated. It could lower the direct monetary cost of maintaining new products and spare part inventories. Furthermore, 3D printing has the potential to reduce the use of hazardous chemicals in manufacturing. It will make disposal of these compounds easier and less expensive and lower the need for their manufacturing.

Green 3D printing develops biodegradable composition as a sustainable FFF composition material based on renewable and natural-origin ingredients [28]. It aids in the alleviation of resource restrictions to promote the spread of socially sustainable supply chain innovation [29].

The advancement in sustainability provides vital opportunities like predictive analytics in 3D printing, environmentally friendly 3D printing, robotics, and 5G technology-based IoT-based cloud manufacturing for future enhancements [30].

5.1.3.2 A Framework of Criteria and Sub-Criteria for Green 3D Printing

---

TABLE 5.2

List of Criteria and Sub-Criteria for Green 3D Printing

Objective: Green 3D Printing

Factors	Content Overview	Source
Materials/equipment used in 3D Printing		
Use of biodegradable plastics	PLA is mainly used biodegradable plastic. Biodegradable plastics can be biodegraded, or it can be recycled. 3D printing filaments for biocompatible and medical purposes are constructed of polymers with low melting temperatures. These materials have been utilized in FDM to make scaffolds and other pieces that merge with human tissues.	[31, 32]
Green material options	Poly hydroxyl alkanets (PHA), polyvinyl alcohol (PVA), polyethylene terephthalate (PET), and high impact polystyrene are the bio-plastics utilized in FDM filaments (HIPS). PHA can be used alone or in combination with PLA. PVA is a biodegradable and water-soluble polymer that is used to support constructions. As a support structure for ABS, Duran printed PVA was used. PVA is printable until 45 minutes after it collects moisture from the air, at which point it becomes impossible to print.	[31, 32]
Refuse to the new influx of plastics	Saying "NO" to the use of non-biodegradable plastics is a better option. The arrival of new kinds of plastics can harm the environment and health more adversely, so stopping the use of new non-biodegradable plastics can help to promote the green environment.	[33]

Green 3D Printing

Objective: Green 3D Printing

Factors	Content Overview	Source
Materials/equipment used in 3D Printing		
Life cycle inventory	The LCI allows you to tax energy use, raw material use, air, water pollution, and total solid waste created (grave being the ultimate is postal). LCI for every material gives the information of usage until its complete disposal. Life cycle analysis of LCI can provide the idea of which long-lasting material can be used for a green environment. This analysis can then be interpreted form of bar graphs, also called as life cycle interpretation.	[34]
Use of light-weight equipment to reduce greenhouse gases	A worldwide problem is reducing greenhouse gas emissions to the point that greenhouse gas concentrations in the atmosphere are stabilized. Using light-weight equipment can be cost efficient and will reduce fires and greenhouse gases.	[18]
Effects on human health		
Reducing exposure to ultrafine particles	Human health safety and exposure to process emissions are also significant during the building phase. Because the plastic is heated to high temperatures, 3D plastic printers produce many ultra-fine particles (UFPs), and the fumes contain harmful by-products. The danger of emitted toxic heavy metal vapor requires attention. The floating metal powder particles could cause respiratory problems in both powder production and AM procedure.	[33]
Using PLA elements in 3Dprinting	Biomaterials like scaffolds are highly useful to improve human health. Bio-plastic filaments like PLA produce fewer fumes and smells than oil-based like ABS. So, using PLA elements in Sprinting is less dangerous to human health and is better for the environment as compared to other plastics.	[9]
Global public safety	If a 3D printer becomes available to every person, people may use it in the wrong manners and can quickly generate weapons; thus, the government must manage the usage of 3D printers up to certain authorities and people.	[35]
Process safety management	Process safety management could be achieved through the use of data collecting methods such as cyber-physical systems and the Internet of Things by avoiding hazards and accidents by identifying and controlling potential sources of failure.	[35]
Process and Usage		
Reduction of weight to 50%	Designers can use topology optimization to calculate the most efficient basics shape of the structure to be cost-efficient. The reduced weight can increase the efficiency of the expenses, especially in the aerospace industry. At the moment, light-weight, high-strength materials with future development are primarily titanium alloy materials, but their high cost hampers their widespread application.	[18, 33]
Usage of carbon fiber reinforced plastics (CFRP)	The carbon composite is a robust and light-weight material for 3D printing CFRPs can be costly to manufacture. Still, they are extensively utilized in aircraft, ship superstructures, automotive, and civil engineering applications where a high strength to weight ratio and stiffness (rigidity) is required.	[33]

(Continued)

TABLE 5.2 (Continued)

Objective: Green 3D Printing

Factors	Content Overview	Source
Materials/equipment used in 3D Printing		
Powder recycling process	The material used during the process is highly dependent on nozzle efficiency. Powder recycling could significantly reduce this load. The unfused powder requires treatments like drying and sieving before reuse due to possible damages to the machine, but they can decrease the extra usage and improve the environment. The metallographic and mechanical properties of Inconel 718 remain the same as brand new powder, and research is needed for other materials.	[33, 36]
AM as a repairing method	Metal AM-based processes could be used, as are pairing tools for both parts and tools repair. This can give a longer life span from technical aspects, which can save a lot of energy and materials.	[13]
Usage of inert gases	From the standpoint of traditional low-CO_2 concrete, OPC (concrete) can be partially replaced by SCMs and inert fillers. Using SCMs to reduce cement consumption is one way to promote sustainable concrete, mainly when SCMs are derived from industrial by-products and recycled wastes.	[9]
Hybrid process	Hybrid manufacturing combines additive processes—3Dprinting, known as additive manufacturing (AM) and subtractive processes, such a smiling. Hybrid manufacturing is that both operations so occur on the same machines, so it is very cost-efficient for manufacturing and production.	[9]
Production costs efficiency		
Using FFF process	It is possible to design fused filament fabrication (FFF) because it is so easy. Printing is growing more popular due to its ease of use and the steady reduction in the cost of printing equipment and materials. An FFF process is a low-cost procedure that can be carried out on an individual basis.	[28]
3D printed manufacturing aid	The number of companies using 3D printing has increased dramatically as the technology has progressed. The tool's applications and use cases vary by industry, but they generally comprise grids, visual and functional prototypes, and even finished products.	.
Schedule Planning	It is process management which includes the activity lists and activity sequencing process. Different processes need different durations. So, schedule planning of operations can improve the time and cost required in the process and can save the electrical consumption.	[9]
3D printer emissions and air quality control for health		
Ventilation of place	The emissions contain ultra-fine aerosols (UFAs) or volatile organic compounds (VOCs), which are harmful to humans. A closed printer design and proper room ventilation are generally suggested to avoid unpleasant odors and associated health hazards.	[9, 37]
Concentration of UFP	The amounts of UFPs produced by 3D printers appear to be similar to those generated by indoor cooking. Still, more research is needed to understand which UFPs 3D printers are referring to assess the health risk. Fans can be employed to deflect fumes, but they may harm the operating temperature and, as a result, the print quality.	[8–10]

Green 3D Printing

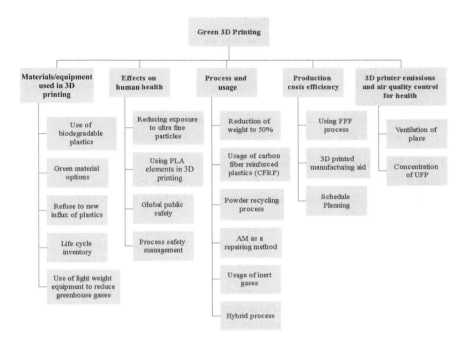

FIGURE 5.3 Framework for green 3D printing.

5.1.4 Applying AHP Method to Our Framework

The analytic hierarchy process (AHP) is a math and brain science-based method for organizing and breaking down complex options. AHP is a standard tool for giving weights to compare other parameters or alternatives. It is based on a simple idea of arbitrary calculation. AHP provides a robust model for problem decision-making, grading, and prioritization, assisting us in handling and developing a model hierarchy depending on our situation. Simple scales can be used to produce comparison measures from surveys or tests of respondents. AHP was used in the following five processes for this barrier rating with India.

- It was created by Thomas L. Saaty in the 1970s and has been refined since then. It is divided into three sections: a clear purpose or issues you seek to communicate, a comprehensive list of possible arrangements (called choices), and assurances that you will decide on the options.
- AHP gives a mandated choice an analytical framework by measuring its criteria and elective alternatives and relates those components to the overall goal. AHP changes the assessments into numbers, which can be contrasted with the entire potential models. This evaluating ability differentiates the AHP from other dynamic strategies
- In the latest of the cycle, mathematical requirements are determined for every elective alternative. These numbers show the most wanted arrangements, according to every input quality.

5.1.4.1 Advantages of AHP Method

i. This method integrates statistical data according to their relations and people's opinions about various factors. From this method, our main aim is to design a framework.
ii. This method combines different sub-methods to resolve disagreement and assessments between every factor decided for the framework.
iii. This method quantitatively manages different types of relations between every factor and subfactors with the help of their weights.

5.1.4.2 Steps for AHP Method

Step 1: Classifications and specific obstacles within each class formed the foundation of the hierarchical structure. The problem was decomposed into a hierarchical tree in this way.

Level 1: Goal/objective (analysis of the main factors for the 3D printing).

Level 2: Represents the priority of the identified five primary elements.

Level 3: Contains overall ranking of priorities of main and sub-factors of the 3D printing.

Step 2: Create a pairwise comparison matrix using Table 5.3's nine-point Saaty scale of relevance. This pairwise dimension and barrier comparisons have been transformed into comparison matrices. To identify the priority matrices, these comparison matrices were solved using the AHP methodology. Barriers within the specific dimensions are coupled with global priority weights.

Step 3: A pairwise comparison matrix was created for goals and barriers in each category based on the survey responses. Comparative judgments were merged while evaluating experts' opinions by applying the geometric mean to the views to generate the matrix of the relative assessment.

Step 4: Determine the consistency. The absolute or relative weights of the greatest Eigenvector are determined. Then, using the equation, determine

TABLE 5.3
Saaty's Scale of Importance for Pairwise Comparison Matrices

Intensity of Importance	Definition	Explanation
1	Equal importance	Two actions contribute equally to the goal.
3	Moderate importance of one over other	One operation has a modest advantage over another based on knowledge and judgment.
5	Essential or strong importance	Activity is strongly favored, and its superiority is shown in practice.
7	Extreme importance	The importance of activity is emphasized, and its superiority is demonstrated in practice.
9	Extreme importance	The evidence that favors one surgery over another is of the highest possible quality.
2, 4, 6, 8	Intermediate values between the two adjacent judgments	When compromise is needed.

TABLE 5.4
Random Consistency Index Values

n	1	2	3	4	5	6	7	8	9	10
RI	0	0	0.58	0.9	1.12	1.24	1.32	1.41	1.45	1.49

the consistency index (CI) value for each n-dimensional matrix (1). The consistency ratio can be determined using an equation based on the CI and random consistency index (RI) (2). The following are the equations:

$$CI = \frac{(\lambda_{max} - n)}{n-1} \quad \ldots\ldots\ldots\ldots \quad (1)$$

The consistency ratio can be calculated as

$$CR = \frac{CI}{RI} \quad \ldots\ldots\ldots\ldots\ldots \quad (2)$$

Table 5.4 displays the value of RI for matrices of order (N) 1–10 generated by approximating random indices using a sample size of 500, where the fluctuation of RI values is dependent on the order of the matrix.

The allowable consistency ratio (CR) range varies depending on the matrix size, for example, 0.05 for a 3x3 matrix, 0.08 for a 4x4 matrix, and 0.1 for all larger matrices, n≥5. If CR is less than or equal to 5, the matrix evaluation is satisfactory or indicates a high level of consistency. However, if CR is greater than the fair value, the matrix is inconsistent, and the evaluation procedure should be evaluated, reassessed, and improved.

5.1.5 DATA ANALYSIS AND RESULTS

A comparative matrix of Table 5.4 was created using the AHP technique in the case of factor prioritizing. Table 5.5 provides the target matrix and consistency ratio for the parameters. We used the same technique for the sub-criteria components.

5.1.6 CONCLUSION

3D printing has two crucial characteristics that make it a "green" technology, according to proponents. To begin with, unlike traditional production techniques such as injection molding, casting, stamping, and cutting, many 3D printing systems produce relatively little waste. The factors that majorly affect 3D printing are:

- Materials/equipment used in 3D printing
- Effects on human health

TABLE 5.5
Aggregate Pairwise Comparison Matrix for Criteria

Main Barriers	F-1	F-2	F-3	F-4	F-5
F-1	1	1.38	1.56	1.67	1.93
F-2	0.77	1	1.32	1.21	1.6
F-3	0.34	0.95	1	1.09	1.21
F-4	0.56	0.52	0.66	1	1.01
F-5	0.22	0.38	0.31	0.51	1

TABLE 5.6
Final Weights for Main and Sub-Criteria Factors for Green 3D Printing

Main Factors	Sub-Factors	Sub-Factors code	Main criteria weight	Weight of Sub Factors	Weight of Overall Levels	Rank
Materials/ equipment used in 3D Printing (F-1)	Use of biodegradable plastics	F-11	0.321	0.444	0.163	1
	Green material options	F-12		0.199	0.073	6
	Refuse to the new influx of plastics	F-13		0.114	0.06	7
	Life cycle inventory	F-14		0.164	0.012	19
	Use of lightweight equipment to reduce greenhouse gases	F-15		0.079	0.013	18
Effects on Human health (F-2)	Reducing exposure to ultra fine particles	F-21	0.312	0.344	0.107	3
	Using PLA elements in 3D printing	F-22		0.387	0.121	2
	Global public safety	F-23		0.145	0.045	8
	Process safety management	F-24		0.125	0.039	9
Process and Usage (F-3)	Reduction of weight to 50%	F-31	0.205	0.366	0.077	5
	Usage of carbon fiber reinforced plastics (CFRP)	F-32		0.167	0.024	14
	Powder recycling process	F-33		0.166	0.014	17
	AM as a repairing method	F-34		0.144	0.03	11
	Usage of inert gases	F-35		0.146	0.03	11
	Hybrid process-	F-36		0.01	0.03	11
Production costs efficiency (F-4)	Using FFF process	F-41	0.116	0.623	0.089	4
	3D printed manufacturing aid	F-42		0.245	0.015	15
	Schedule Planning	F-43		0.132	0.012	19
3D printer emissions and air quality control for health (F-5)	Ventilation of place	F-51	0.046	0.727	0.031	10
	Concentration of UFP	F-52		0.273	0.015	15

Green 3D Printing

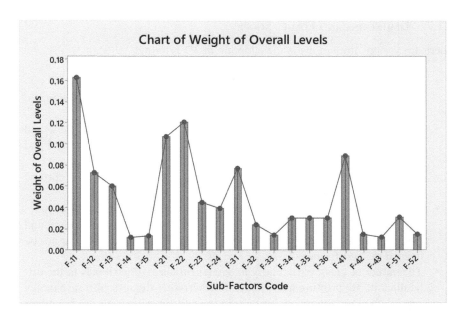

FIGURE 5.4 Overall weights for subfactors.

- Process and usage
- Production costs efficiency
- 3D printer emissions and air quality control for health

AHP is a decision-making technique for complicated situations where several variables or criteria are considered when prioritizing and selecting alternatives. First, we divided our factors and sub-factors into various groups, called experts to define the evaluation criteria, and then decided on the major factors influencing the subfactor selection. At last, we confirmed the selection by giving them the most appropriate numberings. Then, after designing our aspect and subfactors, we established the hierarchy structure based on our factors and subfactors. We computed the element weight of the hierarchies. Then we calculated the eigenvalue and eigenvector for the comparison matrix, and finally, we calculated the relative weights of the element for each level. We got our conclusions with the Delphi-AHP method by using the highest order of relative importance. We observed this conclusion: materials/equipment used in 3D printing (0.321) >effects on human health (0.312) >process and usage (0.205) >production cost efficiency (0.116) > 3D printer emissions and air quality control for health (0.046). In parentheses, we have included the relative weight value results from our analysis. Hence, we concluded that the most influential factor for our framework of green 3D printing is "materials/equipment used in 3D printing" with the relative weight of 0.321. Hence this factor is the most deciding factor from all aspects of this study.

5.1.7 Limitations and Future Scope

There are various limitations in the adoption of 3D printing:

- The lack of well-established methods, standards, and build-material systems are major factors restricting the reliable adoption of 3D printing to manufacture industrial components and products.
- To maintain and improve the awareness of individuals, businesses, and organizations worldwide when making decisions on waste management and socioeconomic demands.
- Nylon-based 3D printing has been found a higher risk of developing chronic interstitial lung disease.
- The final "bio-properties" cannot be determined by using it as a base material. At the filament production stage, adding oils, plasticizers, fillers, and colors decreases biodegradability. As a result, the final product cannot be labeled as "eco-friendly."
- The nature of glass, particularly its great brittleness, contributes to the difficulties of 3D printing glass items. Furthermore, the loss of transparency caused by pores generated during printing and sintering is a significant barrier to the practical implementation of 3D-printed glass objects.

More specifically, a critical view of 3D printing and prioritizing important factors with AHP has been presented. Our insight suggests that more advances in materials/equipment like biomaterials are needed for the success and growth of 3D printing technology in the healthcare and medical industry. Other decision techniques should be used in the future to evaluate emissions from new AM machine types or novel source materials. Furthermore, emissions were very low for most analytes during all AM machine operations; continuing study of emissions and health risks is required to expand and improve technology.

The growth in sustainability gives crucial recommendations such as robotic automation, predictive analytics in 3D printing, eco-friendly 3DP, and 5G technology- and IoT-based cloud manufacturing for future advancements

REFERENCES

[1] Lee, T. C., R. Ramlan, N. Shahrubudin, T. C. Lee, and R. Ramlan. 2019. "An Overview on 3D Printing Technology: Technological, Materials, and Applications." *Procedia Manufacturing* 35: 1286–96. https://doi.org/10.1016/j.promfg.2019.06.089.
[2] Lu, Chuanwei, Chunpeng Wang, Juan Yu, Jifu Wang, and Fuxiang Chu. 2019. "Two-Step 3 D-Printing Approach toward Sustainable, Repairable, Fluorescent Shape-Memory Thermosets Derived from Cellulose and Rosin." *Chemsuschem*, 210037: 1–11. https://doi.org/10.1002/cssc.201902191.
[3] Pyo, Sang-hyun, Pengrui Wang, Wei Zhu, Henry Hwang, John Jamison Warner, and Shaochen Chen. 2016. "Continuous Optical 3D Printing of Green Aliphatic Polyurethanes Continuous Optical 3D Printing of Green Aliphatic Polyurethanes." *ACS Appl. Mater. Interfaces*, 9 (1): 836–844. https://doi.org/10.1021/acsami.6b12500.
[4] Christ, Susanne, Martin Schnabel, Elke Vorndran, Jürgen Groll, and Uwe Gbureck. 2015. "Fiber Reinforcement during 3D Printing." *Materials Letters* 139: 165–68. https://doi.org/10.1016/j.matlet.2014.10.065.

Green 3D Printing

[5] Vora, Hitesh D., and Subrata Sanyal. 2020. *A Comprehensive Review: Metrology in Additive Manufacturing and 3D Printing Technology. Progress in Additive Manufacturing.* Springer International Publishing. https://doi.org/10.1007/s40964-020-00142-6.

[6] Petrick, Irene J., and Timothy W. Simpson. 2013. "3D Printing Disrupts Manufacturing How Economies of One Create New Rules of Competition." *Research-Technology Management*, 56 (6): 12–16, (December). https://doi.org/10.5437/08956308X5606193.

[7] Schubert, Carl, Mark C. Van Langeveld, and Larry A. Donoso. 2014. "Innovations in 3D Printing: A 3D Overview from Optics to Organs." *Br. J. Ophthalmol.* 2014, 98: 159–61. https://doi.org/10.1136/bjophthalmol-2013-304446.

[8] Gopinathan, Janarthanan, and Insup Noh. 2018. "Recent Trends in Bioinks for 3D Printing." *Biomaterials Research* 22 (11). https://doi.org/10.1186/s40824-018-0122-1.

[9] Yan, Qian, Hanhua Dong, Jin Su, Jianhua Han, Bo Song, Qingsong Wei, and Yusheng Shi. 2018. "A Review of 3D Printing Technology for Medical Applications." *Engineering*, 4 (5): 729–742. (July). https://doi.org/10.1016/j.eng.2018.07.021.

[10] Dawood, A., B. Marti Marti, and A. Darwood. 2015. "3D Printing in Dentistry." *Nature Publishing Group* 2. https://doi.org/10.1038/sj.bdj.2015.914.

[11] Ghomi, Erfan Rezvani, Fatemeh Khosravi, Rasoul Esmaeely Neisiany, Sunpreet Singh, and Seeram Ramakrishna. 2020. "Future of additive manufacturing in health care." *Current Opinion in Biomedical Engineering*, 100255. https://doi.org/10.1016/j.cobme.2020.100255.

[12] Ventola, C. Lee. 2014. "Medical Applications for 3D Printing: Current and Projected Uses." *P & T: A Peer-Reviewed Journal for Formulary Management* 39 (10): 704–11.

[13] Lee, Jian-yuan, Jia An, and Chee Kai Chua. 2017. "Fundamentals and Applications of 3D Printing for Novel Materials." *Applied Materials Today* 7: 120–33. https://doi.org/10.1016/j.apmt.2017.02.004.

[14] Joshi, Sunil C., and Abdullah Sheikh. 2016. "3D Printing in Aerospace and Its Long-Term Sustainability 3D printing in Aerospace and Its Long-Term." https://www.researchgate.net/journal/Virtual-and-Physical-Prototyping-1745-2767. *Virtual and Physical Prototyping* 10 (4): 1–11. https://doi.org/10.1080/17452759.2015.1111519.

[15] Wang, Yu-cheng, Toly Chen, and Yung-lan Yeh. 2018. "Advanced 3D Printing Technologies for the Aircraft Industry: A Fuzzy Systematic Approach for Assessing the Critical Factors." *The International Journal of Advanced Manufacturing Technology* 105 (1): 4059–69.

[16] Esperon-miguez, Manuel. 2015. "The Present and Future of Additive Manufacturing in the Aerospace Sector: A Review of Important Aspects." *Proceedings of the Institution of Mechanical Engineers Part G Journal of Aerospace Engineering* 229 (11). (October). https://doi.org/10.1177/0954410014568797.

[17] Sreehitha, V. 2017. "Impact of 3D Printing in Automotive Industries." *International Journal of Mechanical And Production Engineering* 5 (2): 91–94.

[18] Moon, Seung Ki, Yu En Tan, Jihong Hwang, and Yong-jin Yoon. 2014. "Application of 3D Printing Technology for Designing Light-Weight Unmanned Aerial Vehicle Wing Structures." 1 (3): 223–28. https://doi.org/10.1007/s40684-014-0028-x.

[19] Dankar, Iman, Fawaz El Omar, Francesc Sepulcre, and Amira Haddarah. 2021. "Impact of Mechanical and Microstructural Properties of Potato Puree-Food Additive Complexes on Extrusion-Based 3D Printing." *Food and Bioprocess Technology* 11 (2).

[20] Le-bail, Alain, Bianca Chieregato Maniglia, and Patricia Le-bail. 2020. "ScienceDirect Recent Advances and Future Perspective in Additive Manufacturing of Foods Based on 3D Printing." *Current Opinion in Food Science* 35: 54–64. https://doi.org/10.1016/j.cofs.2020.01.009.

[21] Hager, Izabela, Anna Golonka, and Roman Putanowicz. 2016. "3D Printing of Buildings and Building Components as the Future of Sustainable Construction?" *Procedia Engineering* 151: 292–99. https://doi.org/10.1016/j.proeng.2016.07.357.

[22] Khalil, Abdullah, Xiangyu Wang, and Kemal Celik. 2020. "3D Printable Magnesium Oxide Concrete: Towards Sustainable Modern Architecture." *Additive Manufacturing* 33 (November 2019): 101145. https://doi.org/10.1016/j.addma.2020.101145.

[23] Yanar, Numan, Parashuram Kallem, Moon Son, Hosik Park, Seoktae Kang, and Heechul Choi. 2020. "A New Era of Water Treatment Technologies: 3D Printing for Membranes." *Journal of Industrial and Engineering Chemistry* 91: 1–14. https://doi.org/10.1016/j.jiec.2020.07.043.

[24] Zhang, Dao, Xiaofeng Liu, and Jianrong Qiu. 2020. "3D Printing of Glass by Additive Manufacturing Techniques: A Review." *Frontiers of Optoelectronics* 14 (3): 263–77.

[25] Fahmy, Ahmed Raouf, Thomas Becker, and Mario Jekle. 2020. "3D Printing and Additive Manufacturing of Cereal-Based Materials: Quality Analysis of Starch-Based Systems Using a Camera-Based Morphological Approach." *Innovative Food Science and Emerging Technologies* 63 (April): 102384. https://doi.org/10.1016/j.ifset.2020.102384.

[26] Ri, V. L. V., and Dwlrq Lq. 2016. "Tao Peng, Analysis of Energy Utilization in 3D Printing Processes." *Procedia CIRP* 40: 62–67. https://doi.org/10.1016/j.procir.2016.01.055.

[27] Baldassarre, Fabrizio, and Francesca Ricciardi. 2017. "The Additive Manufacturing in the Industry 4.0 Era: The Case of an Italian FabLab." *Journal of Emerging Trends in Marketing and management* 1 (1): 105–15.

[28] Janik, Helena, Maciej Sienkiewicz, and Barbara Mikolaszek. 2021. "PLA – Potato Thermoplastic Starch Filament as a Sustainable Alternative to the Conventional PLA Filament: Processing, Characterization, and FFF 3D Printing." *CS Sustainable Chem. Eng.* 2021, 9 (20): 6923–6938. https://doi.org/10.1021/acssuschemeng.0c09413.

[29] Beltagui, Ahmad, Nathan Kunz, and Stefan Gold. 2019. "The Role of 3D Printing and Open Design on Adoption of Socially Sustainable Supply Chain Innovation." *International Journal of Production Economics* (April 2018): 107462. https://doi.org/10.1016/j.ijpe.2019.07.035.

[30] Singh, Rajesh, Anita Gehlot, Shaik Vaseem Akram, Lovi Raj Gupta, Manoj Kumar Jena, Chander Prakash, Sunpreet Singh, and Raman Kumar. 2021. "Cloud Manufacturing, Internet of Things-Assisted Manufacturing and 3D Printing Technology: Reliable Tools for Sustainable Construction." *Sustainability* 13: 7327. https://doi.org/10.3390/su13137327.

[31] Sanchez, Fabio Cruz, Hakim Boudaoud, Mauricio Camargo, Joshua Pearce, Fabio Cruz Sanchez, Hakim Boudaoud, Mauricio Camargo, and Joshua Pearce. 2020. "Plastic Recycling in Additive Manufacturing: A Systematic Literature Review and Opportunities for the Circular Economy." *Journal of Cleaner Production* 264: 121602.

[32] Daminabo, S. C., S. Goel, S. A. Grammatikos, H. Y. Nezhad, and V. K. Thakur. 2020. "Fused Deposition Modeling-Based Additive Manufacturing (3D Printing): Techniques for Polymer Material Systems." *Materials Today Chemistry* 16: 100248. https://doi.org/10.1016/j.mtchem.2020.100248.

[33] Domingues, J., T. Marques, A. Mateus, P. Carreira, and C. Malça. 2017. "An Additive Manufacturing Solution to Produce Big Green Parts from Tires and Recycled Plastics." *Procedia Manufacturing* 12 (December 2016): 242–48. https://doi.org/10.1016/j.promfg.2017.08.028.

[34] Wittbrodt, B. T., A. G. Glover, J. Laureto, G. C. Anzalone, D. Oppliger, J. L. Irwin, and J. M. Pearce. 2013. "Life-cycle economic analysis of distributed manufacturing with open-source 3-D printers." *Mechatronics* 23 (6): 713–726. https://doi.org/10.1016/j.mechatronics.2013.06.002.

[35] Javidrad, Hamidreza, and Erfan Ahadi. 2018. "Investigation in Environmental and Safety Aspects of Additive Manufacturing (AM)," 5th International Reliability and Safety Engineering Conference, Shiraz, Iran.

[36] Murr, L. E. 2019. "Metallurgy Principles Applied to Powder Bed Fusion 3D Printing/ Additive Manufacturing of Personalized and Optimized Metal and Alloy Biomedical

Implants: An Overview." *Integrative Medicine Research* 9 (1): 1087–103. https://doi.org/10.1016/j.jmrt.2019.12.015.

[37] Zisook, Rachel E., Brooke D. Simmons, Mark Vater, Angela Perez, P. Ellen, Dennis J. Paustenbach, William D. Cyrs, et al. 2020. "Emissions Associated with Operations of Four Different Additive Manufacturing or 3D Printing Technologies." *Journal of Occupational and Environmental Hygiene* 0 (0): 1–16. https://doi.org/10.1080/15459624.2020.1798012.

6 Nanotechnology and Manufacturing

Vijay K. Singh, Puneet Kumar, Manikant Paswan, T. Ch. Anil Kumar, and Saipad B.B.P.J. Sahu

6.1 Introduction 119
6.2 Nanotechnology in Nanomaterial's Manufacturing 121
    6.2.1 Carbon Nanotubes 122
6.3 Nanotechnology in Product-Based Manufacturing 122
6.4 Nanotechnology in Composite Aerospace Structures 124
    6.4.1 Autoclave Cure Prepreg Process 124
    6.4.2 Resin Transfer Moulding Process (RTM) 125
References...126

6.1 INTRODUCTION

No one would have thought of about nanoscale machines or nanites in the twentieth century, however, someone might have heard about them in science fiction. But, this word *nano* has really made revolutionary changes in the thinking of the scientific community about matter or materials in the twenty-first century [1]. A series of scientific breakthroughs have come across in materials and design over that time period, which has pressured manufacturers to think and work at a small scales, possibly less than billionth of a metre. At such small sizes, it is not possible to design and manufacture a device or object with available conventional techniques and technologies. This requirement has brought about a new scientific approach, nanotechnology, which offers boundless opportunities in industrial development and transformation [2, 3].

Nanotechnology is the major force that will navigate the 'fourth industrial revolution' in the modern world [4]. Basically, nanotechnology encompasses the manipulation and scientific understating of material or matter at the molecular and atomic scales. The current status of nanotechnology has opened the door for application in diverse fields such as energy storage, biomedical, manufacturing, food packaging and security. The manufacturing sector is continuous evolving and undergoing revolutionary changes on a regular basis with newly developed technologies. The application of nanotechnology in manufacturing is essentially called atomically precise manufacturing (APM) or nanomanufacturing [5]. It deals with production and application of tiny size material particles, systems, structures and devices, where size lies in the range of 01–100 nanometres. Actually, nanotechnology is changing the way we look at the manufacturing industry and the way we do business, and it is the

DOI: 10.1201/9781003220237-6

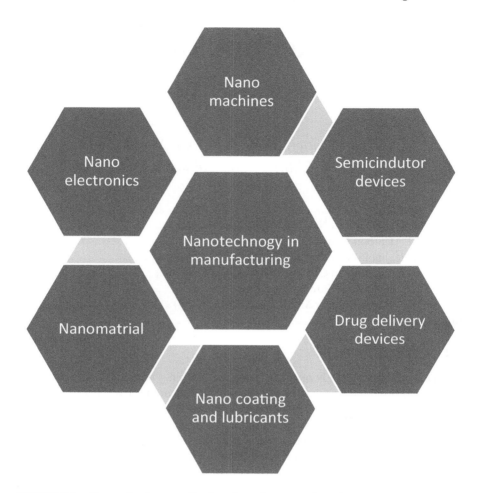

FIGURE 6.1 Nanotechnology applications in various manufacturing sectors.

face of the future. Numerous manufacturing sectors [6, 7] have already started taking advantage of nanotechnology, such as healthcare, automotive, electronics, aerospace, coating and semiconductors, as depicted in Figure 6.1. Material industries are one of the key components of the manufacturing sector, where application of nanotechnology has already existed.

In material industries, nanomaterials and nanoparticle fabrication, in electronic and electric devices, nanoscale transistors, sensors made out of graphene, carbon nanotubes etc. and nanosolar panels in solar energy, in mechanical or robotics devices, nanomachines or nanorobots that operates at nanoscale, in coating industries, nanocoating and nanolubricants for high durability and wear resistance products, in healthcare devices, nanites for drug delivery, hierarchically nanostructured gel to purify water, are the major applications of nanotechnology in manufacturing sector.

6.2 NANOTECHNOLOGY IN NANOMATERIAL'S MANUFACTURING

Manufacturing (or fabrication) and characterization of nanoscale materials require specific capabilities [8]. Because you need to control environmental conditions as well as process parameters, when you deal at nanoscale. Just by introducing some changes in structure at nanoscale, a nanomaterial can be developed with more efficient structure and improved mechanical properties. This could be possible only after nanotechnology came in to existence. Nanomaterials can be sturdier, lighter and safer than existing one and can withstand large pressures and loads. Manufacturers can utilize these nanomaterials for developing more advanced and usable products [9].

Manufacturing of materials, devices, structures and systems at the nanoscale is usually termed as nanomanufacturing. It involves set of industrial processes based on nanotechnology to develop scaled-up, cost effective and reliable products. In real sense, nanomanufacturing performs the integration of manufacturing methodologies with nanoscale material science to deal with multi-domain and multi-scale problems as depicted in Figure 6.2.

Nanomanufacturing [10] utilizes two broad categories of approaches, either bottom-up or top-down, for the manufacturing of nanomaterials and structures. In bottom-up nanomanufacturing, a product is developed by joining the individual molecules or compounds using various combinations of physical and chemical processes.

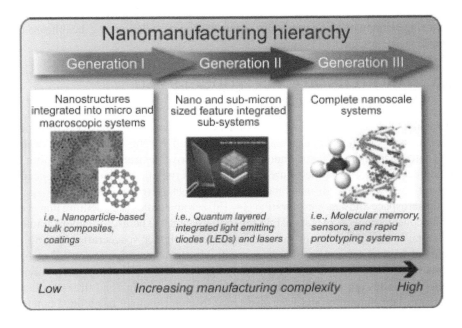

FIGURE 6.2 Evolution of nanomanufacturing.

Source: [10]

Carbon nanotubes that have a unique structure are an excellent example of a bottom-up approach. In a top-down nanomanufacturing approach, a product is developed from larger materials, which break down into nanoscopic components using various physical and chemical processes. There are large range of nanomaterials, which have a wide variety of potential applications in manufacturing of products [11].

6.2.1 CARBON NANOTUBES

One of the most famous and widely used nanoscale materials is the carbon nanotube (CNT). It is the obvious choice of most of current research activities in the nanoscale field due to its unique set of structures and material properties. These tiny materials usually show extra ordinary behavior compared to bulk materials; for example, carbon nanotubes are ultra-strong and lightweight, and have large surface area, and they are more thermally conductive and highly chemically stable. They have numerous applications where a manufacturer requires lightweight, high-wear resistance and large stiffness such as in development of bike frames, industrial robot arms, bullet-proof vests, spaceship components and sailboat hulls [12]. The uniqueness in CNT structure makes them very suitable for drug delivery and water purification. The hexagonal rings in CNT structure plays a major part in filtering out several pollutants from water (biological, chemical, physical, etc.). Sometimes carbon nanofibers are used in the manufacturing of safety wear like bio-textiles due to their antimicrobial properties. CNT is also used in combination with materials of macro size like steel [13] in distributed form to improve its strength.

In addition to nanomaterials, nanotechnology can also be utilized in other manufacturing domains like nanorobotics, nanoelectronics, solar energy, healthcare and nanocoating, as shown in Figure 6.3.

6.3 NANOTECHNOLOGY IN PRODUCT-BASED MANUFACTURING

As nanotechnology is growing continuously, manufacturing industry will experience massive development especially in the automotive, robotics, electronic, coating and aerospace sectors. Nanotechnology will also drive the transformation within the manufacturing sector in the future, which will be crucial, so manufacturers need to be equipped with this advanced technology to unlock its true potential.

In aerospace and space exploration [14–16], aerodynamics is something that needs to be considered as it is responsible for fuel efficiency. The use of nanotechnology in manufacturing has begun to build smaller and lighter spacecraft, reducing the demand for fuel for a operation dramatically. In the automotive sector, all components from tires to chassis are benefited by the use of nanotechnology [17]. Polymer nanocomposites are also being used in the production of high-end tires in order to increase their durability and wear resistance. Nanotechnology has a huge potential to create a revolution in the automotive industry in terms of improved engine efficiency, minimized environmental impact and better electronics systems. Racing car industries have already started reaping the benefits of nanotechnology in design and materials selection.

Nanotechnology and Manufacturing

Solar panel films incorporate nanoparticles to create lightweight, flexible solar cells.

Bamboo-like structure of nitrogen-doped carbon nanotubes for the treatment of cancer

Carbon nanotubes spun into thread

Nanotechnology in Manufacturing

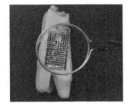
Graphene wireless sensor biotransferred onto the surface of a tooth

Ant-like nanorobots

Finely structured nanoscale coatings

FIGURE 6.3 Nanotechnology based manufacture products.

In electronics, nanotechnology equips the manufacturer to developed highly flexible and tiny devices, which can be used for designing high-performance circuit boards [18]. The application of nanotechnology makes it possible to design highly flexible products such as flexible gas sensors, electric textiles and plastic solar panels. Other areas where nanotechnology has shown a promising impact is in nanomachines or nanites [19]. Right now, these are not widely used, but this is a future technology, where mechanical or robotic devices called *nanorobots* will operate on a nanoscale soon.

In addition, nanotechnology can be applied to develop more efficient nanocoating and lubricants [20], which have various applications in manufacturing industries. These nanocoatings can ensure better wear resistance and durability. Use of nanoscale lubricants ensure the smooth operation and event distribution of loads and makes sure machine components can endure rapid changes in temperature.

Nanotechnology and manufacturing are so well linked that the future development in nanotechnology will certainly be helpful for the manufacturing industry in improving efficiency of operations such as in design, processing, packaging and transportation of goods. This could also help manufacturers in reducing the environmental impact by reducing the use of raw materials, energy and water and reducing the output of greenhouse gases and hazardous waste.

6.4 NANOTECHNOLOGY IN COMPOSITE AEROSPACE STRUCTURES

The effort behind improving the performance of materials based on the concept of reinforcement to the polymer at the nano- and micro level has been ongoing for the last three decades. Apart from the typical structural properties such as strength, stiffness, fracture toughness and energy absorption, there are other important non-structural features like sensing/actuation, self-healing capability, etc.

Multifunctional performance of composites can be understood with an example of enhancement in interlaminar shear strength (ILSS) by 20 percent of epoxy matrix glass fibre composites with the addition of CNTs in small concentration (~0.3 wt. percent) [21]. There are two main paths to make nanocomposites:nano-augmentation (nanoparticles are homogenously dispersed into the matrix of composites) and nano-engineering (introducing pre-organized nanoparticles into the composite laminate).

The novel hybrid composite system may offer properties and functionalities enhancement by reinforcement at the nanolevel as additive in the composites other than obtained from the traditional structural composites. It has been almost two decades since the nanotechnology in fibre reinforced polymer (FRP) was introduced, and a lot of effort has been invested to obtain such materials with outstanding mechanical and electrical properties. The aircraft manufacturing industry has been the origin for most of the composite-related novel developments, and therefore there is always a demand for highly efficient materials which will meet performance requirements in extreme environmental conditions. Nanotechnologies find many potential applications in the space industry, like mechanical toughness and high modulus of damage tolerant structures, ILSS for tubular structures, electrical conductivity for electrical dissipation and lightning strike protection, coefficient of thermal expansion (CTE) for space apertures, protecting the system with severe heat and heat transfer in radiators.

The two most typical technologies in practice to produce aerospace composite structures are prepreg or autoclave and resin transfer moulding (RTM). The first one is the most commonly used technique for manufacturing high-performance and precise aerospace components in the space industry, whereas the second is used in the manufacturing of large complex parts.

6.4.1 AUTOCLAVE CURE PREPREG PROCESS

Prepreg/autoclave technology has been the most commonly used technique for the manufacturing of high-performance and precise components in the aerospace industry, whereas resin transfer techniques have been used as an alternative for the manufacturing of large complex shapes. In the last four decades, prepreg/autoclave technology has been used to assemble a majority of the composite structures in the aerospace industry with the help of prepreg rovings, unidirectional tapes and fabrics. This technology includes various possibilities in the material selection of matrix and fibre phase. Figure 6.4 shows the conventional process of autoclave cured prepreg composite manufacturing. In the very first step of the process, the prepreg fabric/ other fabric is prepared as raw materials and the roll of pre-impregnated fibres. In the next step, the layers as per the required shape and size of the aerospace component

Nanotechnology and Manufacturing 125

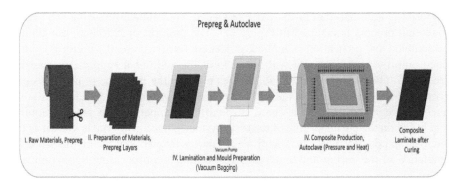

FIGURE 6.4 Process steps in prepreg or autoclave technique for composites manufacturing.
Source: [21]

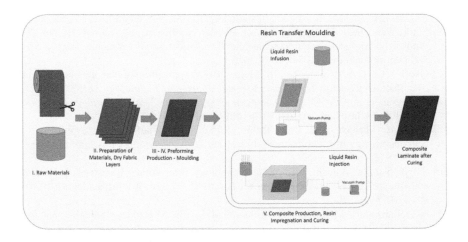

FIGURE 6.5 Process steps in RTM composites manufacturing technique.
Source: [21]

to be produced are cut from the roll. Further, the cut layers need to be laid up manually as per the stacking sequence. Just after the stacking of layers, bagging laminates are placed in a chamber with a vacuum maintained along with the temperature and/or pressure, which controls the resin flow in order to eliminate the air entrapped and other volatiles till the curing of the components. The temperature and pressure are brought to ambient conditions in order to demould the part. In some cases, for high performance polymers the post-curing cycle in a conventional oven is required.

6.4.2 Resin Transfer Moulding Process (RTM)

This is an out-of-autoclave manufacturing process where liquid resin is injected in a dry fibrous preform till the final curing. Here the fibre volume fraction in the final composite is less than that of the parts produced from autoclave curing. The RTM production process is mapped in Figure 6.5. The preform of fibrous reinforcement is assembled and preheating is applied, if needed. Furthermore, the preform is positioned onto the moulding tool with the help of the additional supporting material, if required. The preform impregnation takes place in the presence of pressure and/or vacuum with the help of resin mixture (liquid). As soon as the preform is completely impregnated, the part is cured at room/high temperature as per the requirement based on the curing profile of the resin system. This curing step is known as the composite production process. The produced part is taken out of the mould just after completion of the curing and undergoes post-production processing like trimming/finishing.

REFERENCES

[1] Wennersten, Ronald, Jan Fidler, and Anna Spitsyna. 2008. "Nanotechnology: A New Technological Revolution in the 21st Century." In *Handbook of Performability Engineering*, 943–52. Springer. https://doi.org/10.1007/978-1-84800-131-2_57.

[2] Stark, Wendelin J., Philipp R. Stoessel, Wendel Wohlleben, and Andreas Hafner. 2015. "Industrial Applications of Nanoparticles." *Chemical Society Reviews* 44 (16): 5793–5805. https://doi.org/10.1039/c4cs00362d.

[3] Okoli, J. U., T. A. Briggs, and I. E. Major. 2013. "Application of Nanotechnology in the Manufacturing Sector: A Review." *Nigerian Journal of Technology* 32 (3): 379–85.

[4] Ramsden, Jeremy J. 2013. "The Nanotechnology Industry." *Nanotechnology Perceptions* 9 (2): 102–18. https://doi.org/10.4024/N06RA13A.ntp.09.02.

[5] Mukhtar, Maseeh, and Unni Pillai. 2015. "Nanomanufacturing: Application of Nanotechnology in Manufacturing Industries." *Nanotechnology Law and Business* 12 (1): 5–18.

[6] Singh, Namita, Ashish. 2017. "Nanotechnology Innovations, Industrial Applications and Patents." *Environmental Chemistry Letters* 15 (2): 185–191. https://doi.org/10.1007/s10311-017-0612-8.

[7] Nasrollahzadeh, Mahmoud, S. Mohammad Sajadi, MohaddesehSajjadi, and Zahra Issaabadi. 2019. "An Introduction to Nanotechnology." In *Interface Science and Technology*. Vol. 28, 1–27. Elsevier B.V. https://doi.org/10.1016/B978-0-12-813586-0.00001-8.

[8] Charitidis, Costas A., Pantelitsa Georgiou, Malamatenia A. Koklioti, Aikaterini Flora Trompeta, and Vasileios Markakis. 2014. "Manufacturing Nanomaterials: From Research to Industry." *Manufacturing Review*. EDP Sciences. 1 (2014), 11. https://doi.org/10.1051/mfreview/2014009.

[9] Palit, Sukanchan, and ChaudheryMustansar Hussain. 2020. "Modern Manufacturing and Nanomaterial Perspective." In *Handbook of Nanomaterials for Manufacturing Applications*, 3–20. Elsevier. https://doi.org/10.1016/b978-0-12-821381-0.00001-6.

[10] Grzesik, Wit. 2017. "Nanomanufacturing/Nanotechnology." In *Advanced Machining Processes of Metallic Materials*, 437–65. Elsevier. https://doi.org/10.1016/b978-0-444-63711-6.00017-x.

[11] Hu, Qin, Christopher Tuck, Ricky Wildman, and Richard Hague. 2015. "Application of Nanoparticles in Manufacturing." In *Handbook of Nanoparticles*, 1219–78. Springer International Publishing. https://doi.org/10.1007/978-3-319-15338-4_55.

Nanotechnology and Manufacturing

[12] Santos, Cátia S. C., Barbara Gabriel, MarilysBlanchy, Olivia Menes, Denise García, Miren Blanco, NoemíArconada, and Victor Neto. 2015. "Industrial Applications of Nanoparticles—A Prospective Overview." *Materials Today: Proceedings* 2: 456–65. https://doi.org/10.1016/j.matpr.2015.04.056.

[13] Samareh, Jamshid A., and Emilie J. Siochi. 2017. "Systems Analysis of Carbon Nanotubes: Opportunities and Challenges for Space Applications." *Nanotechnology* 28 (37): 372001. https://doi.org/10.1088/1361-6528/aa7c5a.

[14] Shafique, Muhammad, and Xiaowei Luo. 2019. "Nanotechnology in Transportation Vehicles: An Overview of Its Applications, Environmental, Health and Safety Concerns." *Materials*. MDPI 12 (15): 2493. https://doi.org/10.3390/ma12152493.

[15] Edwards, Eugene, Christina Brantley, and Paul B. Ruffin. 2017. "Overview of Nanotechnology in Military and Aerospace Applications." In *Nanotechnology Commercialization*, 133–76. John Wiley & Sons, Inc. https://doi.org/10.1002/97811193 71762.ch5.

[16] Meyyappan, M., Jessica E. Koehne, and Jin Woo Han. 2015. "Nanoelectronics and Nanosensors for Space Exploration." *MRS Bulletin* 40 (10): 822–28. https://doi.org/10.1557/mrs.2015.223.

[17] Mathew, Jinu, Josny Joy, and Soney C. George. 2019. "Potential Applications of Nanotechnology in Transportation: A Review." *Journal of King Saud University— Science* 31 (4): 586–594. https://doi.org/10.1016/j.jksus.2018.03.015.

[18] Beaumont, Steven P. 1994. "Applications of Nanotechnology in Electronic Devices." In *Conference Proceedings - International Conference on Indium Phosphide and Related Materials*, 371–74. IEEE. https://doi.org/10.1109/iciprm.1994.328247.

[19] Fernando, Raymond H. 2021. "Fernando and Sung." In *Nanotechnology Applications in Coatings ACS Symposium Series*. Vol. 10. American Chemical Society.

[20] Vega Baudrit, José Roberto. 2017. "Nanobots: Development and Future." *International Journal of Biosensors & Bioelectronics* 2 (5). https://doi.org/10.15406/ijbsbe.2017.02.00037.

[21] Kostopoulos, Vassilis, Athanasios Masouras, Athanasios Baltopoulos, Antonios Vavouliotis, George Sotiriadis, and Laurent Pambaguian. 2017. "A Critical Review of Nanotechnologies for Composite Aerospace Structures." *CEAS Space Journal* 9 (1): 35–57. https://doi.org/10.1007/s12567-016-0123-7.

7 Modeling and Optimization of Process Parameters with Single Point Incremental Forming of AA6061 Using Response Surface Method

Assefa Leramo, Devendra Kumar Sinha, and Satyam Shivam Gautam

7.1 Introduction 129
7.2 Experimental Study 131
    7.2.1 Process Description, Material and Method 131
    7.2.2 Design of Experiments 133
    7.2.3 Material Testing Methods...133
7.3 Results and Discussion 134
    7.3.1 Wall Angle..136
    7.3.2 Wall Thinning...139
    7.3.3 Quadratic Equation..140
    7.3.4 Numerical Optimization...142
7.4 Conclusion 143
References...144

7.1 INTRODUCTION

Sheet metal forming, including the incremental forming technique, is gaining prominence among the many metal forming procedures. The industries based on sheet metal forming use various techniques that are mostly dependent on the use of punches and die [1]. Therefore, conservative approaches depending on various punches and die aren't frugally feasible in batch production and at low cost level. To overcome this, incremental sheet metal forming (ISMF) would be a feasible solution for inexpensive sheet metal forming in production sectors of intermittent production with

DOI: 10.1201/9781003220237-7

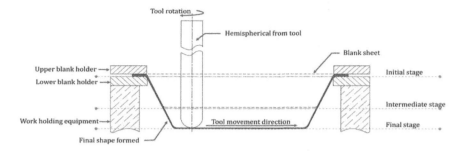

FIGURE 7.1 Schematic representation of Incremental Forming process

applications in automotive, aerospace and various other fields [2–3]. The incremental forming (IF) process twitches due to tool path creation, which is reliant on the shape of the component created. Depending on the movement of tool path a numerical code (NC) is generated that is used to regulate the forming of material in computerized numerical control (CNC) machines. The forming tool moves in the defined tool path over the clamped sheet, thereby creating the necessary shape of the part, as shown in Figure 7.1. Due to down feed movement and frictional contact of tool with the work piece material, local deformation of work piece takes place [4].

Several authors have carried out work on the advancement of forming processes based on incremental forming. Various process factors such as feed rate, spindle speed, tool route, and so on have been studied in order to enhance the essential requirements of the ISMF process [1].

The characteristics of ISMF for AA7075 were investigated experimentally in order to envisage formability. It was found that draw angle of part and depth step affects formability behavior to a higher extent in comparison to tool path [5]. The impact and influence of three alternative tool pathways process factors were investigated [6]. In addition, the process has been analytically and theoretically modeled to study and evaluate the formability behavior. The numerical modeling using LS-DYNA and experimental validation was carried out for analysis of plain strain and biaxial stretching of the fabricated part [7]. The thinning limit of the fabricated part was used to analyze the behavior of formability [8]. It was found that in the fabricated part, the area near to the clamping region of the plate and the area before failure do not satisfy eq (1), which is also known as the sine law, through which thinning of material can be predicted.

$$t_f = t_0 \sin(90 - \varphi) \tag{1}$$

where, t_0 is initial sheet thickness, t_f is the final sheet thickness and φ is the wall angle of the component.

The analysis of deformation behavior with consideration of stress and strain was carried out through development of an analytical model. The deformation of the fabricated part in non-contact zone was also found in order to analyze the accuracy of the part formed [9]. The diagram of forming limit in the case of ISMF process

Modeling and Optimization

appears different than that of conventional forming processes. It was observed that the part with flat regions suffers plain strain stretching, whereas curved regions suffer biaxial stretching [10].

Investigation in the field of ISMF discloses the effect of process characteristics on accuracy and forming limit of the work piece material [11]. An investigation into the thickness control in a new flexible hybrid incremental sheet forming process was carried out. The model was developed for prediction of thickness of work part in terms of process parameters such as drawing depth and performance shape [12]. Experimental and numerical investigations on surface quality for a two-point incremental sheet forming with interpolator was carried out [13]. An investigation on energy consumption during an incremental sheet forming process was carried out through modeling of the effects of process parameters, and the best combinations of process parameters were determined for lowest energy consumption [14].

It has been observed from the literature review that a lot of work has been carried out on ISMF, but little to no investigation has been done on AA6061 for prediction of process characteristics to achieve the best accuracy and surface finish of the fabricated part. Also, interaction effects of the process variables have never been explored. Therefore, in this chapter, the main effect as well as interaction effect of the process parameters such as feed rate (mm/min), step depth (mm), tool diameter (mm) and material thickness (mm) have been investigated on output characteristics such as wall angle (θ) and wall thinning (mm) using RSM. The investigation of wall angle and wall thinning has been conducted experimentally for the fabricated part. The test of significance of the process parameters has been done using ANOVA. Further, best process variables have been proposed for maximum wall angle and wall thinning.

7.2 EXPERIMENTAL STUDY

7.2.1 Process Description, Material and Method

In this investigation, a CNC vertical machining center is utilized for tests (Figure 7.2). A sheet blank of AA6061 material of 150 x 150 x 1 mm (Figure 7.3a) is changed over into a truncated cone with its top face length as well as base length of 20 mm and 100 mm, respectively, and a wall angle of 45°.

Aluminum alloy 6061 is a medium- to high-strength heat-treatable alloy with strength higher than 6005A. It has very good corrosion resistance, formability and very good weldability although reduced strength in the weld zone. It has medium fatigue strength. It has good cold formability in the temper T4, but limited formability in T6 temper. The strain hardening of AA6061 is done in cold rolled form that results in early failure of the parts. Furthermore, the grain size of the sheet is shrinking as a result of this, and the ductility of the sheet has improved. This process is called the Hall-Patch effect [15]. In this manner, to upgrade the formability attributes the aluminum alloy AA6061 clear sheet is annealed and matured at 177°C.

Figure 7.3 shows the 3D model of truncated cone along with the generated spiral tool path using CNC program. The ball end and hemispheric tool form is normally utilized in ISF for sheet metal to obtain high strain rate and cracking delay [16]. Here, a hemispherical form tool of 10 mm diameter (Figure 7.3) is utilized for tests.

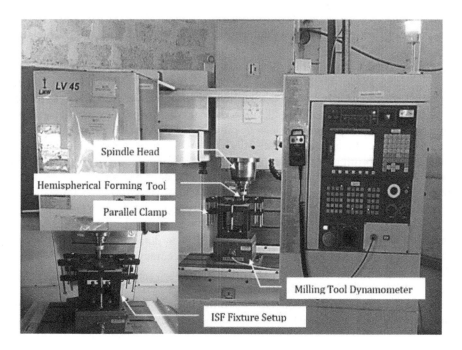

FIGURE 7.2 CNC vertical machining center with incremental forming setup.

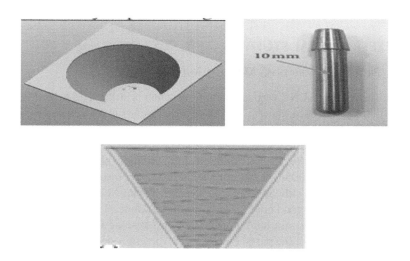

FIGURE 7.3 Modelled truncated cone, hemispherical form tool of diameter 10mm, spiral tool path.

In this investigation, a devoted apparatus arrangement is planned and manufactured for clamping the sheets of measurement 170x170 mm. A plate for backing with an external segment profile is utilized to help the sheet during process of forming. The ISF fixture setups along with its dissembled views are shown in Figure 7.4.

Modeling and Optimization

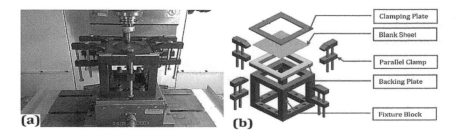

FIGURE 7.4 (a) Fixture setup, (b) exploded view of the fixture.

7.2.2 Design of Experiments

Design experiments are scheduled to investigate the numbers information utilizing arithmetical techniques. Here, RSM-Box Behnken design (BBD) is utilized for experimental design tests. ANOVA is performed to anticipate factors that altogether influence response outputs. The various input factors, such as tool diameter, feed rate, step depth and material thickness, and the corresponding factorial levels of design are presented in Table 7.1. The chosen process parameters for the corresponding maximum wall angle and wall thinning are depicted in Table 7.2 and Table 7.3 respectively. Experimentations are executed as per run order to include the environment clatters with constant rotational speed of the tool at 1000 rpm.

A mix of factors obtained with Box-Behnken design and the response outputs obtained during tests were curve tailored into a quadratic equation, which was additionally utilized for optimization. The regression model of the second order for three-factor plan is presented by equation (2).

$$Y = C_0 + C_1 f + C_2 d + C_3 z + C_4 t + C_5.f.d + C_6.z.t + C_7.f.t. + C_8.d.t \\ + C_9.f^2 + C_{10}.d^2 + C_{11}.t^2 + C_{12}.z^2 \qquad (2)$$

where, f, d, t and z feed rate (mm/min), tool diameter (mm), material thickness (mm), and step depth (mm) as coded factors, C_0, C_1, C_2, C_3, C_4, C_5, C_6, C_7, C_8, C_9, C_{10}, C_{11} and C_{12} are coefficients of regression and Y is the output response predicted.

7.2.3 Material Testing Methods

Tests were directed as per the run order and formed parts are tried for its formability. The important parameters such as maximum wall angle and wall thinning were examined as they straightforwardly add to the formability. The produced part was partitioned into two equal parts and thinning was estimated utilizing the Vernier caliper (Figure 7.5) to assess the thickness in the deformed region of fabricated part. The maximum wall angle is mostly influenced by thinning operation in the formed region. The formed cone slope was utilized for the measurement of wall angle, the segment plane was held perpendicular to z-axis vs. x- axis and the walls in the formed region were identified (Figure 7.5) in four areas for accuracy estimation, taking into account the average value.

TABLE 7.1

Input Process Variables and Output Response Characteristics

Std Order	Run Order	Depth of Step (mm)	Diameter of Tool (mm)	Rate of Feed (mm/min)	Material Thickness (mm)	Maximum Wall Angle (degrees)	Thinning (mm)
2	1	0.2	5	1000	0.8	77.5	0.487
22	2	0.2	5	1000	1	77.2	0.653
13	3	0.2	5	1000	1.2	76.82	0.737
11	4	0.2	10	1500	0.8	77.2	0.489
6	5	0.2	10	1500	1	76.82	0.614
17	6	0.2	10	1500	1.2	75.78	0.747
10	7	0.2	15	2000	0.8	75.4	0.569
14	8	0.2	15	2000	1	75.2	0.628
18	9	0.2	15	2000	1.2	74.2	0.763
15	10	0.5	5	1500	0.8	73.4	0.514
24	11	0.5	5	1500	1	72.45	0.651
16	12	0.5	5	1500	1.2	72.04	0.785
1	13	0.5	10	2000	0.8	73.1	0.516
20	14	0.5	10	2000	1	72.9	0.647
5	15	0.5	10	2000	1.2	71.86	0.787
9	16	0.5	15	1000	0.8	72.3	0.522
26	17	0.5	15	1000	1	72.04	0.654
7	18	0.5	15	1000	1.2	71.04	0.795
23	19	0.8	5	2000	0.8	70.4	0.534
19	20	0.8	5	2000	1	69.4	0.676
12	21	0.8	5	2000	1.2	68.6	0.819
8	22	0.8	10	1000	0.8	69.9	0.538
25	23	0.8	10	1000	1	68.5	0.683
4	24	0.8	10	1000	1.2	66.58	0.839
21	25	0.8	15	1500	0.8	69.6	0.539
27	26	0.8	15	1500	1	69.4	0.676
3	27	0.8	15	1500	1.2	69.2	0.813

7.3 RESULTS AND DISCUSSION

The experimental design was carried out with consideration of input process variables such as rate of feed (mm/min), depth of step (mm), diameter of tool (mm) and material thickness (mm). The output process characteristics such as maximum wall angle (θ) and thickness of wall (mm) are tabulated in Table 7.1. The ANOVA investigation with 95 percent confidence level was carried out for angle and thickness of wall formed (Table 7.2 and 7.3). The determination coefficient (R^2) determined the actual variations for the developed model. The values of R^2 were obtained as 0.9903, and 0.9914 (Tables 7.2 and 7.3) for maximum wall angle and thickness, indicating the model prediction response to be 99.03 percent, and 99.14 percent, respectively. The value of R^2 was kept at 0.9806, and 0.9828, which is closer to its obtained value, and helps in the explanation of output response variations for the model [17–18].

TABLE 7.2
Revised ANOVA Values for Maximum Wall Angle

Source	DF	Adj SS	Adj MS	F-Value	P-Value
Model	13	253.589	19.507	102.17	0.000
z	1	231.412	231.412	1212.07	0.000
d	1	0.344	0.344	1.80	0.202
f	1	0.281	0.281	1.47	0.247
t	1	8.932	8.932	46.79	0.000
z*t	1	0.889	0.889	4.66	0.050
z*f	1	1.940	1.940	10.16	0.007
Error	13	2.482	0.191		
Total	26	256.071			

Model Summary R-Sq = 99.03%; R-Sq (pre)= 94.53%; R-Sq (adj) = 98.06%

TABLE 7.3
Revised ANOVA Values for Wall Thinning

Source	DF	Adj SS	Adj MS	F-Value	P-Value
Model	13	0.329653	0.025358	115.71	0.000
z	1	0.010272	0.010272	46.87	0.000
d	1	0.000008	0.000008	0.04	0.851
f	1	0.000036	0.000036	0.16	0.692
t	1	0.313896	0.313896	1432.33	0.000
z*f	1	0.000800	0.000800	3.65	0.078
z*t	1	0.002080	0.002080	9.49	0.009
Error	13	0.002849	0.000219		
Total	26	0.332502			

Model Summary R-Sq = 99.14% R-Sq (adj) = 98.29% R-Sq (pre) = 95.73%

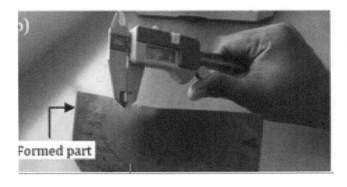

FIGURE 7.5 Wall thickness measurement.

The *p* value of 0.05 of the models as well as the insignificance of lack of fit help to describe the impact of process parameters for the experimental model [19]. The SN ratio reflects the adequate accuracy for the predicted values to that of the average forecast error at design points. Its estimated desired value is ≥ 4 [20]. Here, the adequate accuracies of the model are 102.17 and 115.71 (Tables 7.2 and 7.3), which signifies the efficiency of the model.

7.3.1 Wall Angle

Formability is a critical feature that allows the component to stretch without failure and within dimensional constraints. However, in all conventional processes of forming, the elastic energy storage takes place; formability could be diminished by choosing suitable forming parameters. Wall angle deviation is due to the resilience in the produced section and is directly related with different forming parameters [21]. The wall angle and wall thinning are therefore considered as one of the output responses for the analysis of the formability of the fabricated part. The variance analysis for wall angle is depicted in Table 7.2, and it can be seen that step depth, followed by material thickness, has a major effect on the wall angle. In addition, the feed rate has no impact on the wall angle of the component produced.

It is evident from Figure 7.6 that, when established with maximum step depth, the maximum wall angle achieved is very small. In the deformed area, high elastic energy is stored at increased step depth and the energy also dissipates over time,

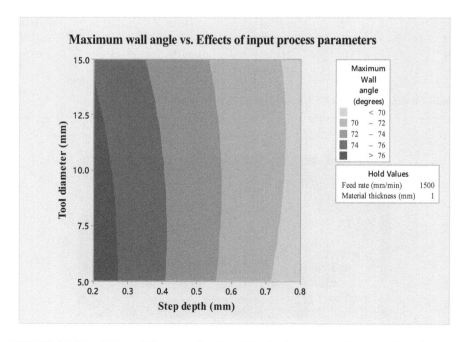

FIGURE 7.6 (A) Effect of diameter of tool and depth of step on maximum wall angle.

Modeling and Optimization

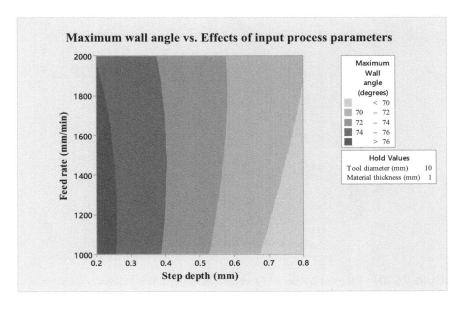

FIGURE 7.6 (B) Effect of rate of feed and depth of step on maximum wall angle.

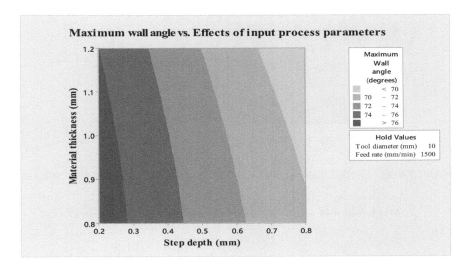

FIGURE 7.6 (C) Effect of thickness of material and depth of step on maximum wall angle.

thus reducing the formability. The higher localized deformation is achieved due to reduced step depth that results in low elastic energy that gets stored in it, causing prolonged formability. The step depth aggravates the wall angle to a greater extent, and is thus acknowledged as a significant element in affecting the maximum wall angle of the shaped component [8, 16].

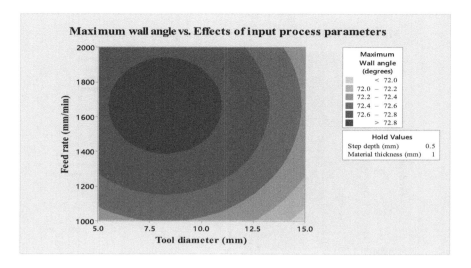

FIGURE 7.6 (D) Effect of diameter of tool and rate of feed on maximum wall angle.

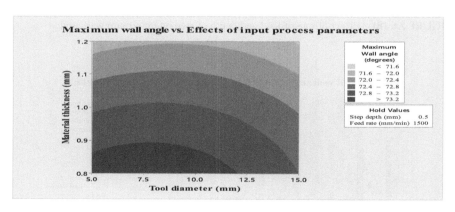

FIGURE 7.6 (E) Effect of diameter of tool and thickness of material on maximum wall angle.

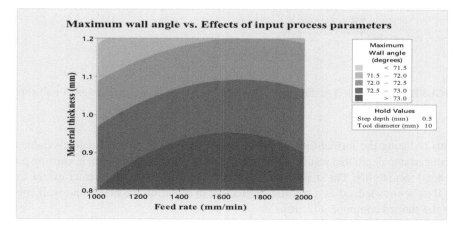

FIGURE 7.6 (F) Effect of thickness of material and rate of feed on maximum wall angle.

7.3.2 Wall Thinning

There is a reduction in the sheet thickness, called thinning, in the forming process. Extreme thickness reduction of the material under the unacceptable limit is considered a failure and, thus, minimum wall thickness found in the shaped section of the truncated cone is used for the analysis of wall thickness. While thinning is unavoidable, by optimizing the process parameters [8–22], it could be reduced within a permissible amount. From ANOVA (Table 7.3), it is shown that the thinning of the formed part is primarily influenced by material thickness, followed by step depth.

The material thinning is found to be maximum when operating with high step depth for thick materials, as shown in Figure 7.7. Higher step depth in the deformation zone causes increased stretching, resulting in reduced wall thickness. Thick materials prohibit the part from strain-hardening by diminishing the time between successive paths of the instrument, thus reducing the thickness of the wall. In addition, stretching is fast under low step depth and material thickness, causing lengthy shear in the tool area, resulting in a reduction in wall thickness [23].

Figure 7.7a shows that step depth has an elevated effect on the component's thinning. The layer is weak to 0.839 mm thickness when operating on a step depth of 0.8 mm. Therefore, premature failure of the part could be prevented with minimal step depth, which also improves the material's formability. The combined impact of feed rate on step depth is seen to be less significant. It is evident from Figure 7.7b that

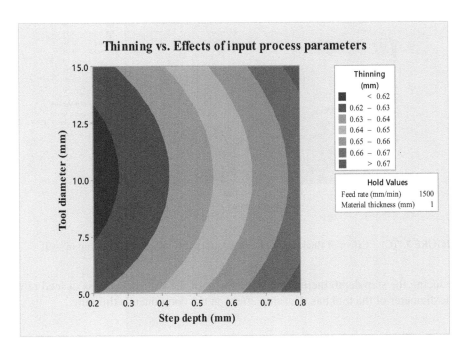

FIGURE 7.7(A) Effect of diameter of tool and depth of step on thinning of wall.

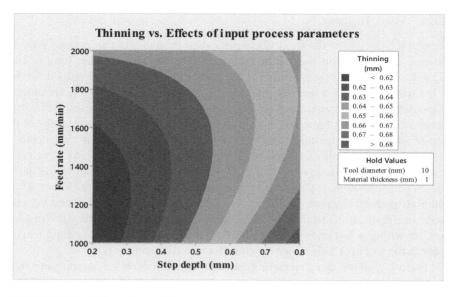

FIGURE 7.7(B) Effect of rate of feed and depth of step on thinning of wall.

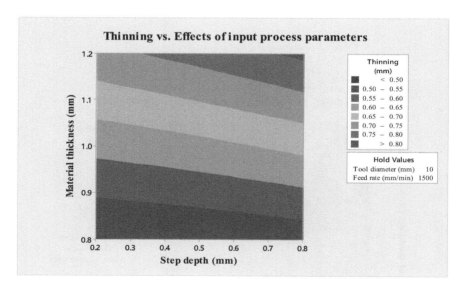

FIGURE 7.7(C) Effect of thickness of material and depth of step on thinning of wall.

reducing the step depth increases the thickness of the wall, regardless of feed rates; the diameter of the tool has the same effect on the component's thinning.

7.3.3 Quadratic Equation

The quadratic equation that is achieved from RSM for predicted maximum wall angle is shown in equation (3).

Modeling and Optimization

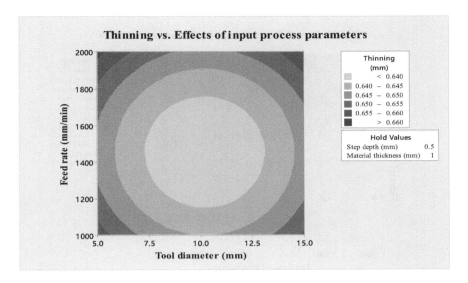

FIGURE 7.7(D) Effect of diameter of tool and rate of feed on thinning of wall.

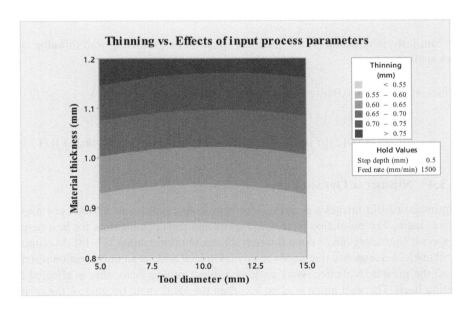

FIGURE 7.7(E) Effect of diameter of tool and thickness of material on thinning of wall.

Maximum wall angle
$$\begin{aligned}
&= 80.77 - 21.76(z) - 0.047(d) + 0.00054(f) + 3.98(t) + 3.36(z^2) \\
&\quad - 0.01111(d^2) - 0.000001(f^2) - 4.03(t^2) + 0.296(z)(d) + 0.00438(z)(f) \\
&\quad - 3.08(z)(t) + 0.082(d)(t) + 0.00085(f)(t)
\end{aligned} \tag{3}$$

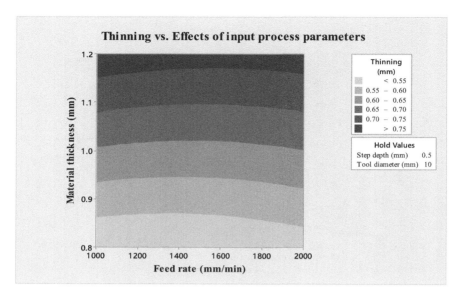

FIGURE 7.7(F) Effect of thickness of material and rate of feed on thinning of wall.

Similarly, in equation (4) the quadratic equation for predicting wall thinning values is given.

$$\begin{aligned}Thinning =\ &0.026-0.048(z)-0.00451(d)+0.000021(f)+0.617(t)\\&+0.0383(z^2)+0.0000478(d^2)+0.040(t^2)+0.00037(z)(d)\\&-0.0000(z)(f)+0.2194(z)(t)-0.00542(d)(t)-0.000062(f)(t)\end{aligned} \quad (4)$$

7.3.4 NUMERICAL OPTIMIZATION

Optimization diminishes a determination with constraints on its variables or maximizes them. The prediction of critical factors and their combination for best output response characteristics is done through statistical optimization [17–18]. Maximum wall angle is a response that needs to be maximized and, as an optimization restriction, the greater the better. Wall angle is a direct profile factor that is affected by spring back. The wall angle is often less than the ideal angle because of the spring back, so optimal parameters are predicted to achieve a greater wall angle. To avoid the failure of the part, the wall thickness must be maximum and hence the maximum wall thickness is a painstaking restraint in the technique of numerical optimization.

According to output process characteristics, the best combination of yield values achieved could be 77.81°, 0.8563mm with a step depth of 0.2mm, diameter of tool 5mm, rate of feed 1000 mm/min and a material thickness of 0.8mm (Figure 7.8) and with a step depth of 0.8mm, diameter of tool 5mm, rate of feed 1000 mm/min and a material thickness of 1.2mm, respectively as shown in (Figure 7.9). Here, the two sets

Modeling and Optimization

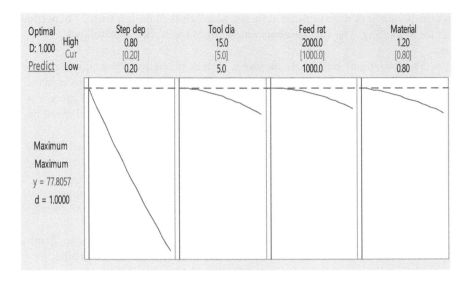

FIGURE 7.8 Optimized response for maximum wall angle and its corresponding parameters.

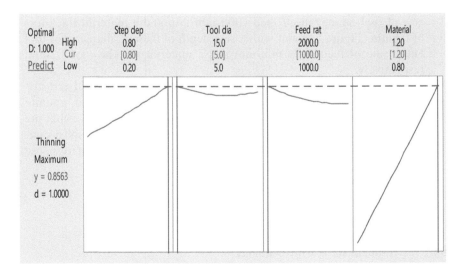

FIGURE 7.9 Optimized response for wall thinning and its corresponding parameters.

of experiments are performed with optimized values of process parameters and the results are 99 percent accurate in prediction of output responses.

7.4 CONCLUSION

In this chapter, the material AA6061, which has a wide range of applications in automotive, aerospace, and medical implants, was tested using incremental forming. The process parameters were optimized by the Box-Behnken design with 27 experiment sets for modelling. The various observations from the work are as follows:

- The maximum wall angle increases with decrease in step depth, material thickness and tool diameter. Consequently, step depth (1212.07) as a factor strongly influences maximum wall angle, followed by material thickness (46.79) compared to the interaction effects step depth (10.16), followed by step depth and tool diameter (4.66) are significant.
- With decrease in depth of step, diameter of tool and thickness of workpiece, the part undergoes deformation and stable stretching, which reduces spring back, and thus the maximum wall angle increases.
- An increase in lubricant viscosity enhances the surface finish, however, it hardly affects the accuracy of profile as well as the thinning of the formed product.
- The F-value of material thickness is (1432.33) followed by step depth (46.87), indicating its significant influence on thinning, relative to other factors. From the interaction effects step depth with value of (9.49), followed by step depth and feed rate (3.65) are significant.
- The numerical optimization of the model suggested that output values achieved could be 77.81°, 0.8563mm with a step depth of 0.2mm, diameter of tool 5mm, rate of feed 1000 mm/min and a material thickness of 0.8mm (Figure 7.9) and with a step depth of 0.8mm, diameter of tool 5mm, rate of feed 1000 mm/min and a material thickness of 1.2mm, respectively.
- The investigation on analysis of energy consumption, force as well as distribution of thickness of the fabricated components can be explored in future work. The investigation can also be extended to polymers and composite materials. The evolutionary approach can be utilized for the global optimization of input process parameters for estimation of output response characteristics.

List of Acronyms

ISF	incremental sheet framing
ISMF	incremental sheet metal forming
RSM	response surface methodology
ANOVA	analysis of variance
IF	incremental forming
NC	numerical control
CNC	computerized numerical control
BBD	Box-Behnken design

REFERENCES

[1] Kim, Y.H., and Park, J.J. 2002. Effect of process parameters on formability in incremental forming of sheet metal. *J. Mater. Process. Technol.* 130: 42–46.

Modeling and Optimization

[2] Emmens, W.C., Sebastiani, G., and van den Boogaard, A.H. 2010. The technology of incremental sheet forming-A brief review of the history. *J. Mater. Process. Technol.* 210: 981–997.

[3] Ambrogio, G., De Napoli, L., Filice, L., Gagliardi, F., and Muzzupappa, M. 2005. Application of incremental forming process for high customized medical product manufacturing. *J. Mater. Process. Technol.*162: 156–162.

[4] Kopac, J., and Kampus, Z. 2005. Incremental sheet metal forming on CNC milling machine-tool. *J. Mater. Process. Technol.* 162: 622–628.

[5] Liu, Z., Li, Y., and Meehan, P.A. 2013. Experimental investigation of mechanical properties, formability and force measurement for AA7075-O aluminum alloy sheets formed by incremental forming. *Int. J. Precis. Eng. Manuf.* 14: 1891–1899.

[6] Rauch, M., Hascoet, J.Y., Hamann, J.C., and Plenel, Y. 2009. Tool path programming optimization for incremental sheet forming applications. *CAD Comput. Aided Des.* 41: 877–885.

[7] Suresh, K., Bagade, S.D., and Regalla, S.P. 2015. Deformation behavior of extra deep drawing steel in single-point incremental forming. *Mater. Manuf. Process.* 30: 1202–1209.

[8] Hussain, G., and Gao, L. 2007. A novel method to test the thinning limits of sheet metals in negative incremental forming. *Int. J. Mach. Tools Manuf.* 47: 419–435.

[9] Fang, Y., Lu, B., Chen, J., Xu, D.K., and Ou, H. 2014. Analytical and experimental investigations on deformation mechanism and fracture behavior in single point incremental forming. *J. Mater. Process. Technol.* 214: 1503–1515.

[10] Park, J.J., and Kim, Y.H. 2003. Fundamental studies on the incremental sheet metal forming technique. *J. Mater. Process. Technol.* 140: 447–453.

[11] Cerro, I., Maidagan, E., Arana, J., Rivero, A., and Rodríguez, P.P. 2006. Theoretical and experimental analysis of the dieless incremental sheet forming process. *J. Mater. Process. Technol.* 177: 404–408.

[12] Zhang, H., Lu, B., Chen, J., Feng, S.L., Li, Z.Q., and Lang, H. 2017. Thickness control in a new flexible hybrid incremental sheet forming process. *J. Eng. Manuf.* 231(5): 779–791.

[13] Li, X., Han, K., Song, Xu., Wang, H., Li, D., Li, Y., and Li, Q. 2020. Investigation on surface quality for two point incremental sheet forming with interpolator. *Chinese J. Aeronaut.* 33(10): 27942806.

[14] Liu, F., Li, X., Li, Y., Wang, Z., Zhai, W., Li, F., and Li, J. 2020. Modeling of the effects of process parameters on energy consumption for incremental sheet forming process. *J. Clean. Prod.* 250: 119456.

[15] Ben Hmida, R., Thibaud, S., Gilbin, A., and Richard, F. 2013. Influence of the initial grain size in single point incremental forming process for thin sheets metal and microparts: Experimental investigations. *Mater. Des.* 45: 155–165.

[16] Oleksik, V. 2014. Influence of geometrical parameters, wall angle and part shape on thickness reduction ofsingle point incremental forming. *Procedia Eng.* 81: 2280–2285.

[17] Mugendiran, V., Gnanavelbabu, A., and Ramadoss, R. 2014. Parameter optimization for surface roughness and wall thickness on AA5052 aluminium alloy by incremental forming using response surface methodology. *Procedia Eng.* 97: 1991, 2000.

[18] Kurra, S., Nasih, H.R., Regalla, S., and Gupta, A.K. 2016. Parametric study and multi-objective optimization in single-point incremental forming of extra deep drawing steel sheets. *Proc. Inst. Mech. Eng., Part B: J. Eng. Manuf.* 230: 825–837.

[19] Dean, A., Morris, M., Stufken, J., and Bingham, D. 2015. *Handbook of design and analysis of experiments.* CRC Press, New York, pp. 300–304.

[20] Davim, J.P. 2015. *Design of experiments in production engineering.* Springer International Publishing, Switzerland, pp. 66–74.

[21] Behera, A.K., Lu, B., and Ou, H. 2016. Characterization of shape and dimensional accuracy of incrementally formed titanium sheet parts with intermediate curvatures between two feature types. *Int. J. Adv. Manuf. Technol.* 83: 1099–1111.

[22] Ambrogio, G., Filice, L., Gagliardi, F., and Micari, F. 2005. Sheet thinning prediction in single point incremental forming. *Adv. Mater. Res.* 6: 479–486.

[23] Jackson, K., and Allwood, J. 2009. The mechanics of incremental sheet forming. *J. Mater. Process. Technol.* 209: 1158–1174.

8 Plasma Fundamentals for Processing of Advanced Materials

Tapan Dash and Bijan Bihari Nayak

8.1 Introduction to Fundamentals of Plasma and Its Classifications 147
    8.1.1 Occurrence of Plasma in Nature 151
    8.1.2 Mundane or Artificially Produced Plasmas 151
8.2 Advantages of Thermal Plasma Processing of Materials 152
8.3 Conclusion 153
References 154

8.1 INTRODUCTION TO FUNDAMENTALS OF PLASMA AND ITS CLASSIFICATIONS

In 1927, the famous scientist I. Langmuir [1] introduced a new term, *plasma*, to the special gas condition in which gases are electrically conducting by the fact that electrons are separated from atoms. Thus, plasma consists of positively charged ions and negatively charged free electrons besides excited and unexcited gaseous atoms and molecules. Although it maintains overall electrical neutrality, its interaction with external electromagnetic fields takes place due to electric fields produced within it. In practice, strict electrical neutrality, i.e. when the number of positive ions is equal to the number of electrons, is hardly possible in plasma. Therefore, plasma is more appropriately considered as an electromagnetic fluid which maintains quasi neutrality. Figure 8.1 illustrates the basic concept of plasma and associated voltage distribution between anode and cathode in the simplest kind of plasma, for example, direct current (dc) glow discharge [2].

Table 8.1 highlights various states of matter and respective densities of species. It may be noted that plasma occupies the fourth position in the table, also currently called the fourth state of matter in physics, and it appears to be so named because of its late discovery, after the three basic states of matter: solid, liquid and gas.

Plasma can be created from individual gas, gaseous mixtures, including noble gases, ambient air or aqueous vapours [3]. Generally, a gas is an electric insulator having a good dielectric constant. A sufficiently large voltage applied across a gap between two electrodes breaks down a gas or gas mixture present in the gap and conducts electricity [4].

DOI: 10.1201/9781003220237-8

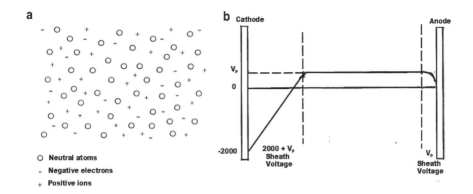

FIGURE 8.1 (a) Conceptual illustration of plasma, (b) voltage distribution across two electrodes.

TABLE 8.1
Various States of Matter and Density of Species

States of matter	Density of species (n) in m^{-3}
Solid	10^{23-25}
Liquid	10^{23-24}
Gas	10^{23}
Plasma, Fourth state	10^{6-34}
Bose-Einstein condensate, Fifth State	10^{23}
Fermionic condensate, Sixth State	10^{23}

This is caused when electrically neutral gas molecules get ionized to form negatively charged electrons and positively charged ions. The voltage required and breakdown behaviour of gas molecules vary with the type of

gas, pressure of gas, flow rate of gas, the materials under treatment including nature of dielectric property, design and arc length of the surfaces across which the voltage is applied, electrode types, high voltage supply nature (e.g. direct current (dc), alternating current (ac), radiofrequency or microwave) and the electrical circuitry design.

[5]

The ionized resulting gas is often known as a discharge or plasma. In plasma the interactions develop among electrically charged particles and with the neutral gas creating special chemical and physical properties. The constituents of the plasma show collective behaviour. Coulombic forces develop among the charges. These forces are affected due to collisions among molecules, ions, etc. In other words, an essential role is played by the *self- generated* electric fields and they decide how the particles move. As a consequence of the collective behaviour, plasma exhibits screen

Plasma Fundamentals

out local density perturbations. The starting gas used to generate/form plasma influences oxidation potential and reaction kinetics of the medium.

Plasmas can be classified on the basis of thermodynamic equilibrium of the gas species such as equilibrium, local equilibrium and non-equilibrium plasma [5–8]. The constituting species of equilibrium plasma are fully ionized and all of them exist at same temperature (usually indicated by ion temperature). The ion temperature is equal to the plasma temperature at this state; it is extremely high in case of astrophysical plasmas (such as sun) [6]. The other two varieties of plasma (local equilibrium and non-equilibrium plasma) are considered as the industrial plasmas in terrestrial conditions [5–8]. The local equilibrium type of plasma known as thermal plasma shows a small fraction of the gas molecules are ionized (not more than 2–3 percent of molecules get dissociated and ionized) in which the heavy species temperature is approximately equal to the electron temperature and the plasma temperature is typically in the range 2,000–25,000 K [9–10] and shows local electrical neutrality, whereas only an upper temperature limit of 2000 K can be achieved by burning fissile fuels [11]. It known that the thermal plasmas are at or close to atmospheric pressure. Thermal plasma has high degree of ionization, high electron density (reaching around $10^{23}–10^{28}$ m^{-3}) and high energy density (10^5-10^6 W/cm^2) [5, 9]. Non-equilibrium plasmas are generally low-pressure (<110 mb) in which electron temperature (kinetic energy of electron is directly converted to temperature, $\varepsilon = 3/2$ kT, where k stands for Boltzmann's constant and T is the temperature, 1 eV ~ 11, 600 K) is several thousands of degrees higher than the temperature of ions, as a result of which plasma shows ambient temperature. We call non-equlibirium plasma [6–7]. They are characterised by their relatively low electron densities < 10^{20} m^{-3} [5]. Different kinds of typical plasmas have different advanced industrial applications [12–14]. Typical kinds of plasmas and corresponding charge density and temperatures are shown in Table 8.2.

Typical properties of plasma include the following:

- Gas or ion temperature (T_g) is generally different from electron temperature (T_e), $T_e"T_g$.
- Obeys Maxwell Boltzmann distribution law.
- More viscous (> 10 times) and conducting (>5 times) than gas.
- In thermal and equilibrium modes of generation, it can produce temperatures from several thousands to millions of degrees K.
- Reaction kinetics is very fast owing to participation of charges.
- Quenching (cooling) rate of the order of million degrees is achievable.
- Produces associated radiation including visible, UV, X-ray, etc.
- Responds to external electromagnetic field due to its self-generated electric field.

In order to define a plasma state from a physics point of view, one has to carefully consider the following parameters [2, 15–16]:

- Charge density (ion, electron) or plasma density (n_i, n_e), neutral density (n_n).
- Plasma potential or space potential (Vp), floating potential (V_f), $V_f < V_p$.

TABLE 8.2
Typical Plasma and Corresponding Charge Densities and Temperatures

Form of Plasma	Charge Density (n/cm³)	Temperature (K)
Light current discharge	10^{11}	10^4
High pr. Heavy current Arc Plasma	10^{15}	10^5
Ionosphere (F-layer)	10^6	$10^{3.5}$
Electrons in metal	10^{23}	$10^{2.5}$
Interstellar space	~0	$10^{3.5}$
Interior of stars (medium magnitude)	10^{27}	$10^{7.5}$
Interior of white dwarf star	10^{32}	$10^{7.5}$
Flame (LPG: Indane/Cal gas)	$<10^3$	1700

- $V_p - V_f = kT/2e \ln(m_i T_e/m_e T_i)$, where m_i = mass of ion, m_e = mass of electron.
  - ion temperature (T_i), electron temperature (T_e), gas temperature (T_g)
- Sheath (the space charge region around a floating substrate with relatively less charge density), $d^2V/dx^2 = -r/e_0$, $dE/dx = r/e_0$
  - where V = potential, E = electric field, r = charge density, permittivity of free space= e_0.
- $n\phi_e/n_e = \exp—e(V_p - V_f)/kT_e$.
- $n\phi_e = n_e \exp(-ej/kT_e)$, where n_e = initial electron density and $n\phi_e$ = electron density in depleted region.

Debye Shielding and Length (λ_D) [2, 15–16]

- Variation of Potential around a Perturbation is derived from Poisson's equation., and given by the equation. d^2V/dx^2" $e^2n_i/kT_e e_0 DV(x)$.
- $DV(x) = DV_0 \exp(- \hat{e}x \, \hat{e}/\lambda_D)$.
- $\lambda_D = (kT_e e_0 /n_e e^2)^{1/2}$.
- Debye Length is the distance over which perturbation is reduced to 0.37(1/e) of its initial value (~ many mm).
- Plasma frequency(W_e), $W_e = (ne^2/m_e e_0)^{1/2}$.

Ambipolar Diffusion [17–18]

- $j_e = -e \, D_e \, dn_e/dx$, where D_e = diffusion of electron, j_e = electric currentdensity due to electrons.
- $j_i = -D_i \, dn_i/dx$, where D_i = diffusion of ion, j_i = electric current density due to ions.
- It implies that under a concentration gradient, electron and ion diffuse at the same rate, i.e. have same diffusion coefficient
- This happens due building up of restraining electric field E.

Plasma Fundamentals

8.1.1 Occurrence of Plasma in Nature

Terrestrial: Stellar & Interstellar

Fire	Sun
Lightening	Stars
Aurora Polaris	Interstellar space
Ionosphere	Solar wind

8.1.2 Mundane or Artificially Produced Plasmas

- Electric arc (8000–10,000 K).
- Welding torch.
- Fluorescent light.
- Plasmatron/Plasma torch (~10,000K).
- Laser plasma (up to million K).
- Artificial sun or TOKOMAK (million K).
- Water jet plasma (50,000K).
- Semiconductor (solid state) plasma.
- Radio frequency (RF)/electron cyclotron resonance (ECR) plasma.
- Quark gluon plasma.
- Nuclear plasma (~million K).

This literature review specially focuses on thermal plasma because it is a widely applied industrial kind of plasma. The devices employed for thermal plasma production occur from a few tens of watts (W) to several megawatts (MW) [10]. It provides rapid evaporation rate and steep temperature gradient and can have wide volume and area of hot zone, high growth rate, etc. [5, 9, 10, 19–20]. For industrial applications it has the most important properties such as (i) the high heat flux or power density 10^5–10^6w/cm^2, which can melt metals and vaporize ceramic particles. These properties of thermal plasma are used in materials and minerals processing [4–5, 9–10, 19–20]. Thermal plasmas have industrial applications

> ranging from small scale (low power thermal plasma devices in industrial operations like arc welding, plasma cutting, plasma spraying, arc lighting, circuit interruption, etc.) to large scale reactors/furnaces (medium and high power devices like electric arc furnaces in the range of hundreds of kW are used for metallurgical applications, mineral processing methods, waste treatment, bio-processing, etc), with some processes typically on an intermediate scale (nanoparticle production, spheroidization, etc).
>
> [4–5, 9–10, 19–21]

Due to the availability of high power density with high temperature in thermal plasma, this plasma has drawn extensive attention, specially for processing of ultra-high temperature ceramics and their composites (WC, TiC, B_4C, SiC, Al_2O_3, AlN, TiN, BN, TiB_2, WB, etc.) in desired shapes and sizes for synthesis of advanced materials and generation of value added materials [5, 22–24]. Thermal plasma technology is very popular for processing new and high-performance advanced materials. Thermal plasma is used in extraction and process metallurgy and covers refining and re-melting of materials, casting operation, etc. [4–5, 9–10, 19–24]. By some new kind of plasma processes deposition of thick nanostructured coatings [4, 9–10, 19–20, 25] could become possible.

In cases of processing of waste into energy, thermal plasma technology has recently been considered as a highly attractive route. It can be easily used for the treatment of various wastes (municipal solid wastes, heavy oil, used car tyres, medical wastes, etc.) for conversion of biomass to syngas at more than 99 percent efficiency (which is mainly composed of H_2, CO and CH_4), because of the ability of the plasma to vaporise anything and destroy any chemical bonds [4, 9, 19–21]. Each of these categories or applications of thermal plasma may include a broad range of endeavours and could be the topic of separate reviews. Most thermal plasmas can be generated by DC or AC electric fields (electric arcs) and inductively coupled RF energy, microwave energy or laser energy [4, 9–10]. Plasma torches are the most commercially available thermal plasma source which are used in industrial uses like welding, melting, furnace and thermal synthesis of materials, etc. Because of these reasons, more representation is required to generalize all the types and applications of thermal plasmas.

8.2 ADVANTAGES OF THERMAL PLASMA PROCESSING OF MATERIALS

Thermal plasma processing technique is relatively simple and provides a cost advantage, particularly in the case of ceramics and refractory compounds that require high-temperature operations. The most generally used gases in thermal plasma are Ar, He, N_2, H_2 and their mixtures [5, 10, 19, 23–24]. Ar is an inert gas, it provides non-oxidizing/non-reducing ambience during processing. Since it is a monoatomic gas, no energy is lost in dissociating molecule and its ionization potential is fairly low (15.759 eV) amongst the inert gases. It is commercially available plentifully unlike other inert gases (e.g. Kr, Ne, He, Xe, etc.). Due to inert nature, electrode service life is more in Ar plasma [19]. Typical cathode materials used in plasma devices are tungsten, thoriated tungsten, graphite, copper, hafnium, niobium and zirconium/zirconia [5, 10, 19, 23–24]. Hollow graphite electrodes are usually used as cathodes, but water-cooled metal electrodes (like thoriated tungsten) have also been used when required. Graphite electrodes are preferred for most processes, as they impose no limits on the current (up to say, 100 kA- scaled up to industrial operation) and its use is simpler (require less skilled maintenance) [26]. The consumable graphite cathode is cheap but has the disadvantage that it can contaminate the product as result of which a more frequent substitution of electrodes is required. Tungsten electrode has to be operated in an inert atmosphere like argon or nitrogen. The non-consumable electrode like tungsten shows a longer life performance but its application is limited because of its erosion behaviour. Copper, graphite and refractory metals possessing excellent thermal and electrical conductivity are generally used as the material for anode [5, 10]. The salient

Plasma Fundamentals

reasons why thermal plasma processing is preferred over conventional processing of materials are summarised as follows:

1. Availability of high temperature (up to 25,000 K) in terrestrial laboratory conditions and large enthalpy (preferred for refractory materials processing).
2. Independent oxygen potential (preferred for reactive materials processing).
3. Very high quenching rate (million K/sec), useful for producing new phases of material from meta stable states.
4. High energy density (10^{5}–10^{6} W/cm^2), suitable for reducing reactor size due to high throughput.
5. Presence of ionic species accelerates oxidation/reduction reaction, thus reaction kinetics of any chemical reaction becomes faster.
6. Ability to feed powders directly into plasma furnace/reactor due to plasma interaction taking place through surface.
7. Environment pollution is minimised due to high temperature involvement in thermal plasma which breaks hazardous and toxic molecules into their constituent molecules of non-toxicity.
8. Acoustic and electrical noise of reactors are reduced to very large extent due to better electrical conductivity of plasma (10–100 ohm^{-1}.cm^{-1}).
9. Purity of higher order in metals is achievable because of higher temperature of reaction in reduction process.

The most important plasma-source devices currently used in processing of materials include "plasma torches and transferred arcs operate in dc mode, inductively coupled (radio frequency (RF)) plasma torches and hybrid combinations of them, etc." [5, 9–10, 19–24]. Generally power sources involved for formation of thermal arc plasma are accompanied by DC power supply [3, 21–24, 26–37]. Polarity reversal takes place in AC, thus causing stoppage of arcing in one half cycle. Particularly for metals and alloys, it therefore causes difficulty to sustain continuous arc in lower-wattage reactors. A DC source generates plasma having the advantage of having control reaction kinetics [19]. Hybrid plasma reactors are therefore designed [38–40] to superpose AC over DC where DC sustains the arc and AC delivers high power so that the resulting combination becomes cheaper. Many reactions which require high temperatures in extractive metallurgy (e.g. carbothermic reduction of metal oxides), which is beyond the scope of conventional furnaces due to thermodynamic behaviour involved and kinetic reaction limitations, can be effectively carried out in plasma furnaces [26].

8.3   CONCLUSION

Plasmas show many outstanding properties which are very useful for processing of materials and their industrial applications. In this chapter we have discussed the fundamental characteristics of plasmas that are important for exploring the properties of plasma science and technology. Research and development work on fundamental plasmas, and plasma processing of materials are now one of the prime areas of focus for researchers. Specially, thermal plasmas have many industrial applications ranging from small-scale applications of arc welding to metallurgical applications, mineral processing methods, waste treatment, bioprocessing, etc. There are many fundaments properties of various plasmas and their applications are yet to be discovered. Over a last few decades, thermal plasmas have been used in industries. Because of its qualities, new application areas are currently being explored. Hence, it is important to explore the details of its fundamental characteristics and applications. There is a need in particular to evaluate the commercial viability of plasma in various industries, which needs be explored in detail to attract the science and technology communities to study the plasma processing of materials.

REFERENCES

[1] I. Langmuir, Oscillations in Ionized Gases, *Proc. Nat. Acad. Sci. U.S.A. (PNAS)*, 14 (1928) 627–637, doi:10.1073/pnas.14.8.627.

[2] J.E. Morris, *ECE 416/516 IC Technologies, Lecture 10: Vacuum and Plasmas*, Springer, 2012.

[3] J.R. Ferrell, A.S. Galov, V.A. Gostev, B.A. Banks, S.P. Weeks, J.A. Fulton, and C.J. Woolverton, "Characterization, Properties and Applications of Nonthermal Plasma: A Novel Pulsed-Based Option." *J. Biotechnol. Biomater.*, 3 (2013) 1000155: 1–10, doi:10.4172/2155-952X.1000155.

[4] S. Shahidi, M. Ghoranneviss, and B. Moazzenchi, *RMUTP International Conference: Textiles & Fashion*, Bangkok and Thailand, July 3–4, 2012.

[5] M.I. Boulos, "Thermal plasma processing." *IEEE Trans. Plasma. Sci.*, 19 (1991) 1078–1089.

[6] Y. Tian, P. Hu, F. Yuan, and A. Hashim (Eds.), *Large-Scale Synthesis of Semiconductor Nanowires by Thermal Plasma, Nanowires -Implementations and Applications, ISBN:978-953-307-318-7*, InTech, Europe and China, 2011.

[7] H-H. Kim, "Nonthermal Plasma Processing for Air-Pollution Control: A Historical Review, Current Issues, and Future Prospects." *Plasma Process. Polym.*, 1 (2004) 91–110.

[8] G. Bonizzoni and E. Vassallo, "Plasma physics and technology; industrial applications." *Vacuum*, 64 (2002) 327–336.

[9] S. Samukawa, M. Hori, S. Rauf, K. Tachibana, P. Bruggeman, G. Kroesen, J.C. Whitehead, A.B. Murphy, A.F. Gutsol, S. Starikovskaia, U. Kortshagen, J.P. Boeuf, T.J. Sommerer, M.J. Kushner, U. Czarnetzki, and N. Mason, "The 2012 plasma roadmap." *J. Phy. D: App. Phys.*, 45 (2012) 253001: 1–37.

[10] N. Venkatramani, "Industrial Plasma Torches and Applications." *Curr. Sci.*, 83 (2002) 254–262.

[11] E. Gomez, D.A. Rani, C.R. Cheeseman, D. Deegan, M. Wise, and A.R. Boccaccini, *J. Hazard. Mater.*, 161 (2009) 614–626.

[12] K.S. Sree Harsha, *Principles of Vapor Deposition of Thin Films*, Elsevier Science, USA, 2005, p. 161.

[13] K. D. Weltmann, J.F. Kolb, M. Holub, D. Uhrlandt, M. Šimek, K. Ostrikov, S. Hamaguchi, U. Cvelbar, M. Černák, B. Locke, A. Fridman, P. Favia, and K. Becker, "The future for plasma science and technology." *Plasma Process Polym.*, 16 (2019) 1–29.

[14] D. Lanbo, Z. Jingsen, Z. Xiuling, W. Hongyang, Li. Hong, Li. Yanqin, and Bu. Decai, "Plasma-Assisted Preparation of Highly Dispersed Cobalt Catalysts for Enhanced Fischer–Tropsch Synthesis Performance." *J. Phys. D Appl. Phys.*, 54 (2021) 333001.

[15] W. Gale, M.Sc. Thesis, *Spacecraft Charge as a Source of Electrical Power for Space Craft*, Air Force Institute of Technology, Wright-Patterson Air Force Base, Ohio, AFIT/GSO/ENP/88D-2, 1988.

[16] J.P. Freidberg, *Plasma Physics and Fusion Energy* (1st ed.), Cambridge University Press, Cambridge, 2008.

[17] Z. Zelinger, P. Kubát, and J. Wild, "The photodissociation of ozone in the Hartley band: A theoretical analysis." *Chem. Phys. Lett.*, 368 (2003) 532–537.

[18] T. V. Losseva and S. I. Popel, "Ambipolar diffusion in complex plasma." *Phys. Rev. E.*, 75 (2007) 046403: 1–6.

[19] J. Heberlein and A.B. Murphy, *J. Phy. D: App. Phys.*, 41 (2008) 053001: 1–20.

[20] M. Mihovsky, "Thermal Plasma Application in Metallurgy (Review)." *J. Univ. Chem. Technol. Metall.*, 45 (2010) 3–18.

[21] T. Watanabe, *21st International Symposium on Plasma Chemistry (ISPC 21)*, Cairns Convention Centre, Queensland, Australia, August 4–9, 2013.

[22] A.V. Samokhin, N.V. Alekseev, S.A. Kornev, M.A. Sinaiskii, Y.V. Blagoveschenskiy, and A.V. Kolesnikov, "Extended characteristics of dispersed composition for nanopowders of plasmachemical synthesis." *Plasma Chem. Plasma Process.*, 33 (2013) 605–616.

[23] T. Dash and B.B. Nayak, "Preparation of WC–W2C Composites by Arc Plasma Melting and Their Characterisations." *Ceram. Int.*, 39 (2013) 3279–3292.

[24] T. Dash and B.B. Nayak, "Preparation and Neutronic Studies of Tungsten Carbide Composite." *Fusion Sci. Technol.*, 65 (2014) 241–247.

[25] R.J. Talib, S. Saad, M.R.M. Toff, and H. Hashim, "Thermal Spray Coating Technology – A Review." *Solid State Sci. Technol.*, 11 (2003) 109–117.

[26] R.T. Jones, T.R. Curr, and N.A. Barcza, *Developments in Plasma Furnace Technology*, Minitek paper No. 8229, Randburg, South Africa, February 4, 1993.

[27] A. Sahu, B.B. Nayak, N. Panigrahi, B.S. Acharya, and B.C. Mohanty, "DC extended arc plasma nitriding of stainless and high carbon steel." *J. Mater. Sci.*, 35 (2000) 71–77.

[28] B.B. Nayak, "Enhancement in the microhardness of arc plasma melted tungsten carbide." *J. Mater. Sci.*, 38 (2003) 2717–2721.

[29] B.C. Mohanty, S.K. Singh, P.K. Mishra, P.K. Sahoo, and S. Adak, Patent no. 1,90,724, Indian, 2003.

[30] B. B. Nayak, T. Dash, and B. K. Mishra, *Proceedings of the 20th Annual Conference on Composites/NanoEngineering*, Beijing, China, July 22–28, 2012.

[31] L. Rao, F. Rivard, and P. Carabin, *4th International Symposium on High-Temperature Metallurgical Processing*, TMS (The Minerals, Metals & Materials Society), 2013, pp. 57–65.

[32] A.N. Klein, R.P. Cardoso, H.C. Pavanati, C. Binder, A.M. Maliska, G. Hammes, D. Fusão, A. Seeber, S.F. Brunatto, and J.L.R. Muzart, "Encyclopedia of Iron, Steel, and Their Alloys." *Plasma Sci. Technol.*, 15 (2013) 70–81.

[33] J.R. Tavares, L. Rao, C. Derboghossian, P. Carabin, A. Kaldas, P. Chevalier, and G. Holcroft, "Large-Scale Plasma Waste Gasification." *IEEE Trans. Plasma Sci.*, 39 (2011) 2908–2909.

[34] B.B. Nayak, T. Dash, and S.K. Pradhan, "Processing and Characterization of Materials." *J. Electron. Spectros. Relat. Phenomena.*, V (2020) 146993: 1–8.

[35] N. Nayak, T. Dash, D. Debasish, B.B. Palei, T.K. Rout, S. Bajpai, and B.B. Nayak, "A novel WC–W2C composite synthesis by arc plasma melt cast technique: microstructural and mechanical studies." *SN Appl. Sci.*, 3 (2021) 1–8.

[36] T. Dash and B.B. Nayak, "A novel WC–W2C composite synthesis by arc plasma melt cast technique: microstructural and mechanical studies." *Ceram. Int.*, 45 (2019) 4771–4780.

[37] T. Dash and B.B. Nayak, "A novel WC–W2C composite synthesis by arc plasma melt cast technique: microstructural and mechanical studies." *Ceram. Int.*, 42 (2016) 445–459.

[38] V. Dembovsky, *Materials Science Monographs, 23: Plasma Metallurgy the Principles*, Elsevier Publication, Amsterdam, Oxford, New York and Tokyo, 1985, pp. 222–225.

[39] A. Bogaerts and E.C. Neyts, "Plasma Technology: An Emerging Technology for Energy Storage." *ACS Energy Lett.*, 3 (2018) 1013–1027.

[40] A. Bogaerts, X. Tu, J.C. Whitehead, et al., "The 2020 plasma catalysis roadmap." *J. Phys. D: Appl. Phys.*, 53 (2020) 443001: 1–51.

9 The Concept of Rotary Ultrasonic Bone Machining during Orthopaedic Surgeries

Raj Agarwal, Jaskaran Singh, Vishal Gupta, and Ravinder Pal Singh

9.1 Introduction 157
9.2 Application of Rotary Ultrasonic-Assisted Machining in the Biomedical Field 160
9.3 Effect of Input Parameters on Output Response 164
    9.3.1 Effect of Rotational Speed on Output Response 164
    9.3.2 Effect of Feed Rate on Output Response..165
    9.3.3 Effect of Drill Design on Output Response.......................................168
9.4 Limitation of the Study and Future Scope...169
9.5 Conclusion 169
References...169

9.1 INTRODUCTION

Bone is the infrastructure of the human body, it provides a rigid framework that offers protection and support, as well as attachment sites for muscles that are essential for locomotion and movement [1]. Bones are living, metabolically active and growing soft fibrous material that has dense connective tissues. The 206 distinct bones in the human body come in four basic shapes: long bone, short bone, flat bone and irregular bone. Bone tissue is made of distinct types of bone cells. The properties of bone are hard, brittle and have poor thermal conductivity [2]. It is a composite tissue having a hard outer layer made of crystalline calcium phosphate (hydroxyapatite) with small amounts of other mineral substances, covering a soft spongy structure made of the protein collagen. The outer layer gives strength and the inner honeycomb-like structure containing a matrix gives the flexibility required by the body. The exterior outer layer is known as cortical bone and the inner spongy surface is known as cancellous bone [3]. With the task of structurally supporting the body, 80 percent of the weight of a human skeleton is due to cortical bones, and they are very dense compared to cancellous bones. The bones are stiffer and stronger at higher strain rates, which is a way to compensate for the higher loads and stresses imposed by vigorous activity or

DOI: 10.1201/9781003220237-9

super physiological loading. Both cortical and cancellous bone contains living cells, which help make repairs when bones are broken or injured.

As the bone helps to strengthen the body, so the mechanical behaviour of bone is exceedingly important for estimating fracture risks. Some part of a body is constantly under compressive loads, which may increase the risk of fracture. The fracturing of bone is a common issue because of sports injuries, age, osteoporosis, industrial and transport accidents, etc. At the time of fracture, bones break, or bone alignments may be dislocated. A bone fracture occurs at the point where the physical force exerted on bone is much stronger than the bone itself. It is a medical condition where the continuity of the bone is broken partially or completely. The treatment of bone fracture usually involves restoring fractured bone to its initial position and immobilizing it until restoration and reconstruction take place. There are two main methods for the treatment of bone fracture: direct and indirect. The indirect method is a time-consuming process, during which the body part is immobilized from the outside and its movement is relatively restricted. For outer fixing of bone, plaster casts, body casts and orthopaedic casts are used, as depicted in Figure 9.1a. There are a few disadvantages with the indirect method, like wrong bone alignment and a lengthy process of healing; the indirect method is also known as the *conventional* method. To overcome these limitations, a direct method has been introduced by orthopaedic surgeons where the fractured bone is treated internally and directly by surgical procedure to reposition and secure the bones. The direct method implies the fixation of the fractured part of bone by exposing the skin and fixing the fractured bone with the help of implants and screws, as illustrated in Figure 9.1b. Bone drilling is performed to create holes at the right position on fractured sites and an implant is fixed. In this approach, immobilization is done by fixation of bone with implants and screws.

Bone drilling removes the misalignment problem of bone but produces a new issue, i.e. thermal narcosis, which means completely or partially damaging nearby parts (soft tissues) of bone which cannot be regenerated again once burned. During drilling of bone, heat is generated due to which thermal burning of soft tissues occurs which may result in the failure of bone generation and screw implant failure.

Hillery and Shuaib noticed no significant difference in the temperature when drilling with distinct point angles of 70°, 80° and 90° with a 23° positive rake angle [4]. This research was carried out to look into the matter of the most suitable drill

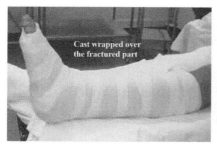 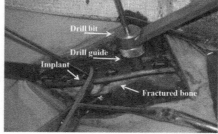

FIGURE 9.1 Methods to treat a fractured bone: (a) indirect method, (b) direct method.

Rotary Ultrasonic Bone Machining

shape and optimum drilling speed. From experimentation, it was revealed that as the drilling depth increases during drilling, the temperature increases. Figure 9.2 shows a fractured bone undergoing femur surgery with the direct approach of bone drilling. In Figure 9.2, a screw is inserted in implants of cortical bone at the left side and a cannulated screw is inserted in the cancellous bone at the top. This method of treatment is more secure and stable than the conventional indirect approach [4]. Therefore, many medical surgeries, for example trauma, dental, orthopaedic and neurosurgical operations, require drilling on bone.

The purpose of bone drilling is to provide support to the fractured part of bone directly that helps heal the bone fracture faster [5]. The orthopaedic drilling of bone is very similar to mechanical drilling processes which results in increasing the cutting force, temperature at the surrounding bone material. As the fractured bone heals it causes generation of microcracks on the surface of bone. The holding power at the bone-screw joint depends on anchorage strength and pull-out strength between screw

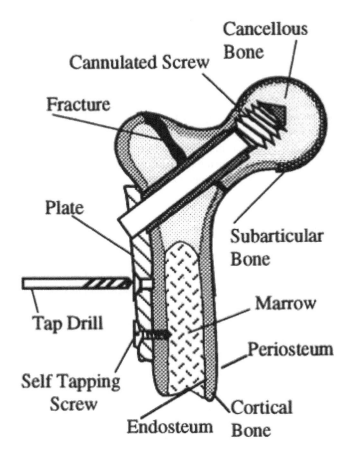

FIGURE 9.2 Drilling on bone for surgery.

Source: Reprinted with permission from Elsevier [4].

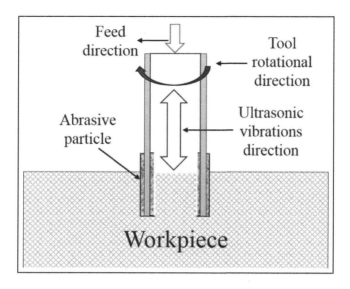

FIGURE 9.3 Illustration of the RUAM process.

and bone. Damaged bones with microcracks are notable to hold the screws and implants for a very long period leading to implant failure. Moreover, the excessive increase in force, torque and temperature lead to thermal necrosis, poor pull-out strength of the screw, and screw loosening that may lead to intolerable pain at the fracture site and affect the reliability of surgery causing reoperation and revision surgery [6–7].

In the last few years, various advanced machining technologies were invented and introduced with the aim of reducing surgical complications and eliminating obstructions for the surgeon. The rotary ultrasonic-assisted machining (RUAM) is the most explored technique for eliminating the problem of escalated temperature, enhanced cutting forces, torque and reducing microcracks in the surrounding bone surface. RUAM is proven to be a better alternative as compared to conventional machining during bone surgeries. In this technique, ultrasonic vibrations are provided to the drilling tool that produce intermittent (non-uniform) contact between drilling tool and bone surface, as illustrated in Figure 9.3. The interruption caused by ultrasonic vibrations in the drilling tool discontinues the direct continuous friction at bone-tool interface. This technique has been successfully implemented and showed improvements during surgery by reducing temperature rise [8], enhancing pull-out strength of screw [9], reducing microcracks [10], minimizing the cutting forces and torque [11–12] and reducing tool wear [13]. The various studies related to RUAM are discussed in the following section.

9.2 APPLICATION OF ROTARY ULTRASONIC-ASSISTED MACHINING IN THE BIOMEDICAL FIELD

Today, researchers have discovered a new drilling process, Rotary Ultrasonic Bone Drilling (RUBD). Ultrasonically assisted drilling for cutting bones and fixing implants are becoming popular in orthopaedic, neuro and dental surgeries. It is one of the advanced technologies that have been developed to overcome most of limitations of conventional drilling. In this method, an ultrasonic machine is used for tool vibration with the help of a transducer at an ultrasonic frequency of 20–40 kHz. With the help of a piezoelectric transducer an ultrasonic vibration is given to the attached tool on the horn. When high power electricity is passed from them, the grains get scattered in their specific direction continuously [14–17]. The ultrasonic power is supplied from the ultrasonic generator to the piezoelectric actuators—the element which generates ultrasonic vibration. Because the tool performs rotary as well as longitudinal motion, there is no continuous contact between bone and the tool which produces less cutting force generation and less temperature leads to better surface texture and fewer microcracks [14, 18–20]. It provides precise cutting and enhances the take-off strength of the cortical screw as compared to conventional bone drilling (CBD) [9–10]. Various authors used the novel technique of RUBD to improve surgical procedures and minimize the failure of osteosynthesis. Some of the studies are discussed and reviewed as follows:

> Gupta et al. [6] compared RUBD with CBD on porcine bone for force and torque measurement and suggested that for lower force and torque, the recommended process parameters setting is high rotational speed with low feed rate. Moreover, it was concluded that ultrasonic drilling of bone generates a lower temperature than conventional drilling of bone because at higher amplitude, time of contact between tool and bone decreases.

FIGURE 9.4 Experimental setup of (a) RUBD, (b) CBD; 1) CNC collet; 2) carbon brushes; 3) slip rings; 4) collar; 5) horn; 6) nut and collet; 7) hollow tool; 8) holding fixture for bone; 9) bone sample; 10) conventional surgical drill bit.

Source: No permission is required according to the journal policy [6].

Singh et al. [7] compared ultrasonic drilling with conventional drilling for bovine bone to observe surface morphology. It was concluded that better finish and better quality holes are observed with ultrasonic machining due to less mechanical forces and that it gives a finer surface of the bone specimen as compared to conventional drilling. Moreover, the ultrasonic drilling of bone produces less bone debris compared to conventional bone drilling.

Chen et al. [8] used low frequency vibrational assisted drilling to monitor thermal injuries induced during drilling of bovine bone. It was observed that vibrational assisted drilling effectively reduces temperature rise as this process interrupts the continuous accumulation of cutting heat.

Agarwal et al. [9] used a rotary ultrasonic drilling process on porcine bone to enhance the pull-out strength of the cortical screw. It was witnessed that rotary ultrasonic drilling produces a precise cut that enhances the bone healing and improves the holding power at the bone-screw interface due to which pull-out strength of the cortical screw increases.

Gupta and Pandey [14] used rotary ultrasonic drilling (RUD) of pig bone for controlling temperature rise. They used diamond coated hollow tools for drilling bones. Thermocouples were used for measuring the temperature changes of bone. Effect of RUD parameters, which are rotational speed, feed rate, amplitude, and drill diameter on change in temperature at different drilling location at a time of bone drilling was monitored. It was concluded that as compared to the RUD of bone, the conventional drilling method generates a higher temperature, which is very useful for orthopaedic drilling operations. This is because temperature increases with rise in rotational speed, feed rate and diameter of drill, but decreases with a rise in vibrational amplitude.

Shakouri et al. [15] have compared conventional and ultrasonic-assisted drilling of bovine bone to observe thermal necrosis during the drilling process. It was observed that the ultrasonic-assisted drilling produces less thermal necrosis in comparison to conventional bone drilling.

Singh et al. [16] introduced a new surgical vibrational bone drilling technique. During vibration drilling of bone, the twisted tool of the drill vibrates along longitudinal direction in a specific frequency and revolves around its axis with a constant feed rate. Bovine bone was used for this investigation. For heat generation the major process parameter was rotational speed. Its contribution on the result was 80.53 percent. At minimum revolving speed of (600 r/min), osteonecrosis was very low, and at maximum revolving speed of (3000 r/min), osteonecrosis was high. Hence, according to results, vibrational drilling technique showed less damage as compared to conventional drilling specimens.

Gupta et al. [19] compared RUD to conventional drilling of porcine bone to optimize force, torque and temperature. It was observed that RUD has the capability to reduce temperature, force and torque in comparison with conventional drilling.

Singh et al. [21] compared RUD with conventional drilling of human bones for monitoring the change in temperature and cutting force. It was also observed that the cutting force generated was 30–40 percent lower and temperature rise was 50–55 percent lower during rotary ultrasonic drilling compared with conventional drilling of bone.

Wang et al. [22] investigated the microcrack analysis observed at the drill site during the vibrational drilling and conventional drilling of bovine bone. It was concluded that vibrational drilling of bone reduces the generation of microcracks that affect postoperative recovery.

Alam et al. [23] compared conventional drilling with ultrasonic drilling of bovine bone for optimizing the surface roughness of drilled bone specimens. It was concluded that the drilling done by the ultrasonic process produces a finer and smoother surface finish of the hole as compared to the drilling done by conventional processes.

Wang et al. [24] performed a study on bovine cortical bone to compare the occurrence of microcrack levels in conventional drilling (CD) and vibrational (VD) processes. Their experimental results of using VD gave fewer and shorter microcracks as compared to CD. Shorter microcracks mean less fatigue damage and less stress fracture. It was also concluded that the temperature falls with the rise in the level of vibration frequency and amplitude.

Alam et al. [25] compared ultrasonically assisted drilling (UAD) and conventional drilling (CD) of bovine bone to monitor thrust force and torque. To minimize the thrust force produced during drilling, the UAD and CD techniques were compared with variations in different input parameters. Higher cutting speed induces greater material removal rate which increases the production rate. With a different set of drilling parameters such as drilling speed (600–3000 rpm), drill diameter (4mm), drill cutting edge angle (65 degrees), federate (10–50 mm/min) for both CD and UAD, the thrust force and torque was monitored. Vibration amplitude (5–25) and vibration frequency (10–30) were only provided during the UAD drilling technique. UAD was more advantageous as compared to CD in terms of improved chip removal rates from drilling site and decreases in force generation.

Alam and Silberschmidt [26] analyzed the temperature rise during ultrasonic-assisted drilling and conventional drilling of bovine bone. It was observed that ultrasonic-assisted drilling minimized the rise in temperature as compared to the drilling of bone by conventional processes.

Gupta et al. [27] used conventional bone drilling of porcine bone and various drilling parameters for studying temperature variation during drilling. They used diamond coated hollow tools with grit size 80–100 (very coarse) to compare it with a conventional twisted drill bit. Various parameters included rotational speed, feed rate and outer diameter. According to their conclusions a diamond coated hollow tool with very coarse abrasive grit size minimizes the temperature rise compared to a surgical twisted drill bit. There was a major difference between the temperature of a twisted drill bit (70.8°C) and a diamond coated hollow drill bit (48.2°C).

Agarwal et al. [28] used rotary ultrasonic drilling with a novel designed hollow drill bit with different abrasive grits sizes to observe the efficacy of bone graft. It was observed that rotary ultrasonic drilling can provide solid

9.3 EFFECT OF INPUT PARAMETERS ON OUTPUT RESPONSE

Most orthopaedic surgeries done using bone drilling are based on fixing plates and implants by the cortical screws. Healing of bone and sustainability of joints will depend upon the drilling parameters, i.e. rotational speed, drill diameter, feed rate, drilling force, temperature generated, microcracks in bone, etc. The most significant process parameters reported can be broadly categorized into two major groups, which are drilling parameters and drill specifications. The detailed information of process parameters that affect bone drilling are mentioned in Table 9.1.

Li et al. [30] investigated the heat generation during drilling of bone by establishing a finite element method (FEM)-based three-dimensional model focusing on feed rate, spindle speed and drill diameter. They revealed that if any of three parameters rose, it would increase the drilling temperature. Then they used an empirical formula to predict temperature rise and to optimize the parameters. They also introduced the idea of intermittent feed drilling to be more effective in reducing thermal necrosis of bone tissue.

Singh et al. [31] optimized process parameters for better quality characteristics of the hole made during bone drilling processes. Three different types of drill bits such as abrasive coated drill bit, twisted drill bit and hollow drill bit without abrasives were compared to drill the bovine cortical bone. It was observed that better surrounding surface can be obtained from the twisted drill bit as compared with abrasive and hollowed drill bits.

9.3.1 Effect of Rotational Speed on Output Response

It is observed from the literature that rotational speed is one of the most important parameters during drilling that highly influence the fractured bone during the surgery. The rotational speed during drilling refers to the rate at which the drilling tool

TABLE 9.1

Major Process Parameters Affecting Bone Drilling

Serial no.	Drilling parameters	Drill specifications
1	Drilling rotational speed	Diameter of drill bit
2	Feed rate	Cutting face
3	Cooling	Drill point
4	Internal cooling and External cooling	Rake angle
5	Drilling depth	Clearance and helix angle
6	Drilling energy	Point angle and Chisel edge
7	Predrilling	Drill wear

Source: [29]

Rotary Ultrasonic Bone Machining

rotates. It is measured in revolutions per minute (RPM), which refers to the number of rotations that the drill makes in one minute. RPM speeds are crucial for successful drilling during surgery. Most of the research studies the impact of rotational speed on temperature rise, cohered microcracks, torque and force generation.

Gupta et al. [32] reported the effect of rotating speed and feed rate on chip formation during drilling of porcine bone. It was concluded that at high speed, forces are reduced and large sized chips are formed. The reason is that at low rotational speeds, there is high a coefficient of friction between bone and the tool interface. At higher feed rates, more material is removed in a specific period as compared to the lower feed rate.

Singh et al. [33] reported the impact of rotating speed and feed rate on bovine bone drilling to monitor thermal necrosis. It was demonstrated that temperature could be reduced far below the critical values if an optimized set of drill parameters were implanted. A thermocouple actuator was used for measuring the change in temperature of bone. It was concluded that rise in bone temperature is inversely proportional to the drilling speed and feed rate.

Gupta et al. [6] varied rotational speed and feed rate for investigating the microcracks generated during drilling of porcine bone. It was concluded that higher rotational speed of drilling results in smaller width in microcracks, as depicted in Figure 9.5. The rise in the microcrack length of bone results in less strength in the bone-screw joint. The poor surface finish of bone enhances the increase in microcracks, which are the main reason for decreases in elastic modulus and stiffness of bone. In addition, if the length of cracks in bone is equal to or greater than 300 μm, then these microcracks result in bone failure, but below 300 μm, the bone can be repaired.

9.3.2 Effect of Feed Rate on Output Response

Feed rate is another most substantial process parameter that is used during the drilling of bone. The drill feed is the rate that a drill bit is inserted into the bone workpiece in order to create a precise hole. The drill feed rate can be controlled by a surgeon and it is highly dependent on the bone mass or bone mineral density of fractured bone and the design and geometries of the drilling tool. Most of the authors have studied the impact of feed rate on output responses.

Sui and Sugita [34] studied the effect of different geometries of drill bit, rotational speed and feed rate on the thrust force and torque induced during the drilling of bovine and swine femur bone. It was observed that geometries of drill bit have a significant impact on drilling force. Moreover, with rise in a rotational speed, torque and thrust force can be minimized.

Wang et al. [35] conducted conventional a drilling experiment on bovine bone by varying parameters like rotational speed, feed rate and drill-bit diameter. According to the results, feed rate is a very important parameter that had to be controlled to avoid breaking of the bone. In addition, feed rate is directly proportional to forces generated. Also, the chip formation process changes with higher speeds, which might affect the bone as the friction coefficient decreases. In addition, tool diameter is responsible for an increase in force and torque during bone drilling. Because of increments in tool diameter, the cross-section area between bone and the tool changes.

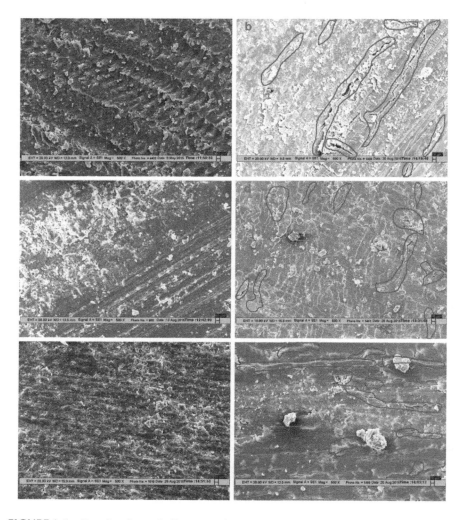

FIGURE 9.5 Rotational speed effects on microcracks generation: (a), (c), (e) RUBD; (b), (d), (f) conventional surgical bone drilling (CSBD) group. (a), (b) 500 rpm; (c), (d) 1500 rpm; (e), (f) 2500 rpm

Source: No permission is required according to the journal policy [6].

Pandey et al. [36] conducted experiments on bovine bone for optimizing drilling parameters, i.e. spindle speed and feed rate in view of minimizing the force and temperature generated during drilling. It was noted that feed rate has the highest contribution (81.94 percent) to the composite desirability. The investigation with a feed rate of 40mm/min and rotational speed of 500 RPM had the highest performance index and these were the recommended process parameters for minimum temperature settings.

Singh et al. [37] varied rotational speed, feed rate and drill diameter during the drilling of human bone for monitoring the change in temperature. It was found that the temperature increases with rise in drill diameter and revolving speed. Furthermore, the temperature rise can be reduced by increasing the feed rate.

Gupta et al. [6] varied rotational speed and feed rate for investigating cortical screw pull-out strength at bone-screw interface. It was concluded the cortical screw's axial pull-out strength increases with higher rotating speed and the strength of the screw decreases with the increase in feed rate. The impact of feed rate on the microcracks can be seen in Figure 9.6. It was witnessed that increasing feed rate increases

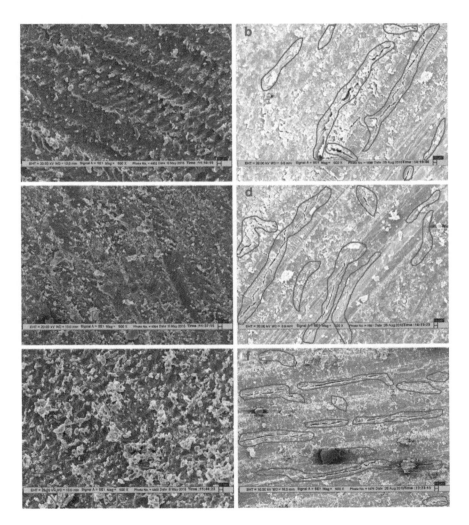

FIGURE 9.6 Feed rate effect on microcracks generation: (a), (c), (e) RUBD; (b), (d), (f) CSBD. (a), (b) 10 mm/min; (c), (d) 30 mm/min; (e), (f) 50 mm/min.

Source: No permission is required according to the journal policy [6].

the microcracks that occur in a conventional drilling process as compared with ultrasonic assisted drilling.

9.3.3 Effect of Drill Design on Output Response

The drill bit is the most important tool in the creation of holes for screw insertion to provide rigid internal fixations at fractured sites. Various studies from the last few decades show the impact of drill design and drill-bit geometries on the output responses.

Singh et al. [17] studied the impact of rotating speed, feed rate and tool type during the drilling of bovine bone. It was observed that tool type plays a major role in the surface morphology at drill site. It was witnessed that a twisted drill bit produces a precise cut at adrill site.

Udiljak et al. [38] studied the effect of drill bit geometry on temperature rise and axial drilling force. This study showed that cutting speed statistically has no significant effect on axial drilling force, but affects the temperature rise. The authors conducted some experiments with standard surgical drills and two-phase drills, and they found that with the two-phase drills there was a minimal increase in bone temperature.

Agarwal et al. [28] compared surgical twisted drill bits with hollow drill bits to examine the graft quality of bone observed during the drilling process of porcine bone. It was observed that the hollow designed tool is capable of providing cylindrical bone grafts as compared to the surgical twisted drill bit.

Singh et al. [39] studied the impact of various process parameters like rotational speed, feed rate and type of tool on bovine bone drilling to monitor the surface morphology of drilled holes. It was observed that a rotational speed of 2000 RPM and a feed rate of 50 mm/min is suitable for optimized surface morphology. Moreover, the twisted drill bit produces fine and circular holes as compared to coated and hollow drill bits, as witnessed in Figure 9.7. The chip morphology can be reduced with the use of fine-diamond-impregnated drill bits [40–41].

FIGURE 9.7 Drilled hole surface morphology produced with (a) abrasive coated tool,(b) hollow pipe tool and (c) twisted drill tool.

Source: Reprinted with permission from Elsevier [39].

9.4 LIMITATION OF THE STUDY AND FUTURE SCOPE

The experimental investigations were performed on the cadaver bone of porcine femur, however, the structural integrity of bone varies in different species of animal. In addition, the drilling was performed in an in-vitro environment and bone specimens were clamped in the especially designed fixture. In the future, the potential of RUBD could be explored on different species of animals in in-vivo environments. In addition, a histopathological study could be performed to demonstrate the in-situ effect of high temperatures on osteocytes.

9.5 CONCLUSION

Bone fractures are very common, and hence there needs to be better care while drilling bone to avoid the problems associated with bone drilling, such as temperature rise, high rates of microcracks, surface morphology, delamination, increase in cutting force and torque during the surgery. Therefore, advanced machining technology called rotary ultrasonic machining has been invented in recent years to enhance orthopaedic surgeries. This novel technique produces interruptions during drilling that provides non-uniform contact between bone-tool interfaces. This drilling technique offers a number of benefits, including reducing temperature rise, reducing cutting force and induced torque, minimizing the microcracks at drill site, improving the holding strength of bone screws at bone-screw interfaces, enhancing the pull-out strength of bone screws, decreasing interlayer delamination and diminishing implant failure rates that reduce the chances of reoperation and revision surgery. The advantage of this technology has been successfully validated over the last few years by various researchers. This technique of drilling is a game-changing technology that offers a number of advantages over standard conventional bone drilling. This technique is currently not implemented in clinical applications, but the future of rotary ultrasonic machining during orthopaedic surgeries is very bright.

REFERENCES

[1] F. Johannesdottir and M. L. Bouxsein, "Bone structure and biomechanics," in *Encyclopedia of Endocrine Diseases*, Elsevier, 2018, vol. 4, pp. 392–400.

[2] J. A. Woo, L. Y. Savio, R. E. Debski, J. Zeminski, S. D. Abramowitch, M. S. Chan Saw, S. Serena, and S. Fenwick, "Injury and repair of ligaments and tendons," *Annual Review of Biomedical Engineering*, vol. 2, no. 1, pp. 83–118, 2000.

[3] M. F. A. Akhbar and A. W. Sulong, "Surgical drill bit design and thermomechanical damage in bone drilling: A review," *Annals of Biomedical Engineering*, pp. 1–28, 2020.

[4] M. T. Hillery and I. Shuaib, "Temperature effects in the drilling of human and bovine bone," *Journal of Materials Processing Technology*, vol. 92, pp. 302–308, 1999.

[5] M. Jamil, S. Rafique, A. M. Khan, H. Hegab, M. Mia, and M. K. Gupta, "Comprehensive analysis on orthopedic drilling: A state-of-the-art review," *Proceedings of the Institution of Mechanical Engineers, Part H: Journal of Engineering in Medicine*, vol. 234, no. 6, pp. 537–561, 2020.

[6] V. Gupta, P. M. Pandey, and V. V Silberschmidt, "Rotary ultrasonic bone drilling: Improved pullout strength and reduced damage," *Medical Engineering and Physics*, vol. 41, pp. 1–8, 2017.

[7] G. Singh, V. Jain, and D. Gupta, "Comparative study for surface topography of bone drilling using conventional drilling and loose abrasive machining," *Proceedings of the Institution of Mechanical Engineers, Part H: Journal of Engineering in Medicine*, vol. 229, no. 3, pp. 225–231, 2015.

[8] J. Chen, Q. An, F. Zou, D. Yu, and M. Chen, "Analysis of low-frequency vibration-assisted bone drilling in reducing thermal injury," *Materials and Manufacturing Processes*, vol. 36, no. 1, pp. 27–38, 2021.

[9] R. Agarwal, V. Jain, V. Gupta, S. Saxena, and V. Dwibedi, "Effect of surface topography on pull-out strength of cortical screw after ultrasonic bone drilling: An in vitro study," *Journal of the Brazilian Society of Mechanical Sciences and Engineering*, vol. 42, no. 7, pp. 1–13, 2020.

[10] A. R. Singh, Ravinder Pal Gupta, Vishal Pandey, and Pulak Mohan Mridha, "Effect of drilling techniques on microcracks and pull-out strength of cortical screw fixed in human tibia: An in-vitro study," *Annals of Biomedical Engineering*, pp. 1–12, 2020.

[11] R. P. Singh, P. M. Pandey, A. R. Mridha, and T. Joshi, "Experimental investigations and statistical modeling of cutting force and torque in rotary ultrasonic bone drilling of human cadaver bone," *Proceedings of the Institution of Mechanical Engineers, Part H: Journal of Engineering in Medicine*, vol. 234, no. 2, pp. 148–162, 2020.

[12] R. P. Singh and P. M. Pandey, "Comparison of conventional and ultrasonic drilling on cutting force in porcine and human femur," in *Proceedings of the ASME 2020 15th International Manufacturing Science and Engineering Conference*. American Society of Mechanical Engineers, 2020, pp. 1–8.

[13] V. Gupta and P. M. Pandey, "In-situ tool wear monitoring and its effects on the performance of porcine cortical bone drilling: A comparative in-vitro investigation," *Mechanics of Advanced Materials and Modern Processes*, vol. 3, no. 1, pp. 1–7, 2017.

[14] V. Gupta and P. M. Pandey, "Experimental investigation and statistical modeling of temperature rise in rotary ultrasonic bone drilling," *Medical Engineering and Physics*, vol. 38, no. 11, pp. 1330–1338, 2016.

[15] M. Shakouri, Ehsan Sadeghi, Mohammad H. Karafi, Mohammad R. Maerefat, and Mehdi Farzin, "An in vitro study of thermal necrosis in ultrasonic-assisted drilling of bone," *Proceedings of the Institution of Mechanical Engineers, Part H: Journal of Engineering in Medicine*, vol. 229, no. 2, pp. 137–149, 2015.

[16] G. Singh, V. Jain, D. Gupta, and A. Sharma, "Parametric effect of vibrational drilling on osteonecrosis and comparative histopathology study with conventional drilling of cortical bone," *Proceedings of the Institution of Mechanical Engineers, Part H: Journal of Engineering in Medicine*, vol. 232, no. 10, pp. 975–986, 2018.

[17] G. Singh, R. Jindal, V. Jain, and D. Gupta, "Effect of tool and drilling parameters on surface topography of bone drilled holes: An in vitro study," *2016 International Conference for Students on Applied Engineering (ICSAE)*, 2017, pp. 196–200.

[18] V. Gupta, P. M. Pandey, R. K. Gupta, and A. R. Mridha, "Rotary ultrasonic drilling on bone : A novel technique to put an end to thermal injury to bone," *Proceedings of the Institution of Mechanical Engineers, Part H: Journal of Engineering in Medicine*, vol. 231, no. 3, pp. 189–196, 2017.

[19] V. Gupta, R. P. Singh, P. M. Pandey, and R. Gupta, "In vitro comparison of conventional surgical and rotary ultrasonic bone drilling techniques," *Proceedings of the Institution of Mechanical Engineers, Part H: Journal of Engineering in Medicine*, vol. 234, no. 4, pp. 398–411, 2020.

[20] J. Mago, R. Kumar, R. Agrawal, A. Singh, and V. Srivastava, "Modeling of linear shrinkage in PLA parts fabricated by 3D printing using TOPSIS method," in *Advances*

in Additive Manufacturing and Joining, M. Shunmugam, M. S. and Kanthababu, eds. Springer Singapore, 2018, pp. 267–276.

[21] R. P. Singh, P. M. Pandey, and C. Behera, "Effects of rotary ultrasonic bone drilling on cutting force and temperature in the human bones," *Proceedings of the Institution of Mechanical Engineers, Part H: Journal of Engineering in Medicine*, vol. 234, no. 8, pp. 829–842, 2020.

[22] Y. Wang, M. Cao, Y. Zhao, G. Zhou, W. Liu, and D. Li, "Experimental investigations on microcracks in vibrational and conventional drilling of cortical bone," *Journal of Nanomaterials*, vol. 5, p. 845205, 2013.

[23] K. Alam, A. V Mitrofanov, and V. V Silberschmidt, "Measurements of surface roughness in conventional and ultrasonically assisted bone drilling," *American Journal of Biomedical Sciences*, vol. 1, no. 4, pp. 312–320, 2009.

[24] Y. Wang *et al.*, "Experimental investigations and finite element simulation of cutting heat in vibrational and conventional drilling of cortical bone," *Medical Engineering and Physics*, vol. 36, no. 11, pp. 1408–1415, 2014.

[25] K. Alam, A. V Mitrofanov, and V. V Silberschmidt, "Experimental investigations of forces and torque in conventional and ultrasonically-assisted drilling of cortical bone," *Medical Engineering and Physics*, vol. 33, no. 2, pp. 234–239, 2011.

[26] K. Alam and V. V. Silberschmidt, "Analysis of temperature in conventional and ultrasonically-assisted drilling of cortical bone with infrared thermography," *Technology and Health Care*, vol. 22, no. 2, pp. 243–252, 2014.

[27] V. Gupta, P. M. Pandey, A. R. Mridha, and R. K. Gupta, "Effect of various parameters on the temperature distribution in conventional and diamond coated hollow tool bone drilling: A comparative study," *Procedia Engineering*, vol. 184, pp. 90–98, 2017.

[28] R. Agarwal, V. Gupta, and V. Jain, "A novel technique of harvesting cortical bone grafts during orthopaedic surgeries," *Journal of the Brazilian Society of Mechanical Sciences and Engineering*, vol. 8, no. 43, pp. 1–14, 2021.

[29] R. K. Pandey and S. S. Panda, "Drilling of bone: A comprehensive review," *Journal of Clinical Orthopaedics and Trauma*, vol. 4, no. 1, pp. 15–30, 2013.

[30] X. Li, W. Zhu, J. Wang, and Y. Deng, "Optimization of bone drilling process based on finite element analysis," *Applied Thermal Engineering*, vol. 108, pp. 211–220, 2016.

[31] G. Singh, V. Jain, and D. Gupta, "Multi-objective performance investigation of orthopaedic bone drilling using Taguchi membership function," *Proceedings of the Institution of Mechanical Engineers, Part H: Journal of Engineering in Medicine*, vol. 231, no. 12, pp. 1133–1139, 2017.

[32] V. Gupta and P. M. Pandey, "An in-vitro study of cutting force and torque during rotary ultrasonic bone drilling," *Proceedings of the Institution of Mechanical Engineers, Part B: Journal of Engineering Manufacture*, vol. 232, no. 9, pp. 1549–1560, 2016.

[33] G. D. Singh, G. Gahi, and A. Jain, "An investigation on thermal necrosis during bone drilling," *International Journal of Machining and Machinability of Materials*, vol. 18, no. 4, pp. 341–349, 2016.

[34] J. Sui and N. Sugita, "Experimental study of thrust force and torque for drilling cortical bone," *Annals of Biomedical Engineering*, vol. 47, no. 3, pp. 802–812, 2019.

[35] W. Wang, Y. Shi, N. Yang, and X. Yuan, "Experimental analysis of drilling process in cortical bone," *Medical Engineering and Physics*, vol. 36, no. 2, pp. 261–266, 2014.

[36] R. Kumar and S. S. Panda, "Optimization of bone drilling process with multiple performance characteristics using desirability analysis," *Procedia—Social and Behavioral Sciences*, vol. 9, no. Icbee 2013, pp. 48–53, 2014.

[37] R. Pal, P. Mohan, and A. Ranjan, "An in-vitro study of temperature rise during rotary ultrasonic bone drilling of human bone," *Medical Engineering and Physics*, vol. 79, pp. 33–43, 2020.

[38] Sja. Udiljak, Toma Ciglar, and Damir Skoric, "Investigation into bone drilling and thermal bone necrosis," *Advances in Production Engineering & Management*, vol. 2, no. 3, pp. 103–112, 2007.

[39] G. Singh, V. Jain, D. Gupta, and A. Ghai, "Optimization of process parameters for drilled hole quality characteristics during cortical bone drilling using Taguchi method," *Journal of the Mechanical Behavior of Biomedical Materials*, vol. 62, pp. 355–365, 2016.

[40] R. Agarwal, V. Gupta, and J. Singh, "Additive manufacturing-based design approaches and challenges for orthopaedic bone screws: A state-of-the-art review," *Journal of the Brazilian Society of Mechanical Sciences and Engineering*, vol. 44, no. 1, pp. 1–25, 2022.

[41] R. Agarwal, V. Gupta, and J. Singh, "A novel drill bit design for reducing bone-chipmorphology in orthopaedic bone drilling," *Materials Today Proceedings*, vol. 52, no. 1, pp. 2–7, 2022.

10 Advancement in Magnetic Field Assisted Finishing Processes

Girish C. Verma, Dayanidhi K. Pathak, Pawan Sharma, Aviral Misra, and Pulak M. Pandey

10.1 Introduction 173
10.2 Basics of MFAF Process... 175
    10.2.1 Mechanism of MFAF Processes.. 175
    10.2.2 Process Parameters in the MFAF Process............................. 176
10.3 Hybrid Magnetic Field Assisted Finishing Processes 178
    10.3.1 Chemo-Mechanical Magnetorheological Finishing 178
    10.3.2 Ultrasonic Assisted Magnetic Abrasive Finishing 179
    10.3.3 Electrolyte Magnetic Abrasive Finishing 179
10.4 Different MFAF Processes for Non-Planar Surfaces 181
    10.4.1 Magnetorheological Finishing Process 181
    10.4.2 Magnetorheological Abrasive Flow Finishing 182
    10.4.3 Rotary Magnetorheological Abrasive Flow Finishing Process 184
    10.4.4 Magnetic Float Polishing.. 184
    10.4.5 Ball End Magnetorheological Polishing Process 185
    10.4.6 Magnetorheological Jet Finishing Process 187
    10.4.7 Magnetorheological Abrasive Honing Process.................... 187
    10.4.8 Magnetic Abrasive Finishing of Cylindrical Parts............. 188
    10.4.9 Internal Magnetic Abrasive Finishing 189
10.5 Conclusions... 190
10.6 Limitations and Future Scope of MFAF Process 191
    10.6.1 Limitations.. 191
    10.6.2 Future Scope 191
References.. 191

10.1 INTRODUCTION

In the present era, dimensional accuracy and surface quality are the two most stringent requirements of manufacturing industries. Many of these products have freeform surfaces due to their functional and aesthetic requirements, like metallic mirrors, medical implants, etc. Additionally, due to the dependency of fatigue strength on surface roughness, it also becomes a major concern for the product's life. For improving the surface finish, different conventional finishing techniques like grinding and

DOI: 10.1201/9781003220237-10

lapping are available; however, most of these processes cannot finish freeform surfaces. Also, applying these conventional finishing processes induces residual stresses in the finished surface, resulting in variation in its mechanical properties. Due to these limitations, non-conventional finishing processes are used, such as chemical polishing, magnetic field assisted finishing (MFAF), etc. Among these, MFAF processes are considered versatile because of their ability to finish non-conductive material [1–2]. Additionally, the polishing pressure in MFAF processes can be regulated precisely, resulting in a smoother and more accurately finished surface.

In recent years, many MFAF processes, for example, the cylindrical magnetic abrasive finishing (CMAF) process, ball end magneto rheological process (BEMRP), rotational magnetorheological abrasive flow finishing (RMRAFF), magnetic float polishing (MFP), etc. have been developed. All these processes were developed to finish different geometrical shapes [3] for example, BEMRP is utilized to finish 3D surfaces, while, to finish spherical balls, MFP is employed, etc. [4]. The finishing in all mentioned processes is carried out by a flexible magnetic abrasive brush (FMAB).

All the MFAF processes use abrasive media (abrasive and carbonyl iron particle mixture) in the external magnetic field for executing the finishing operation. Mainly aluminium oxide (Al_2O_3) or silicon carbide (SiC) were more frequently used as abrasive particles [6–7]. The bonded or unbonded form of the mixture of abrasive particles and ferromagnetic material might be used. For preparing the bonded mixture, the abrasive particles and iron particles are sintered together in an inert environment and further followed by ball milling and sieving. The experimental results have shown that both the bonded and/or unbonded abrasive particles result in effective finishing of surface [8–9].

However, due to the complexity of preparing the bonded abrasive particle, most researchers have used unbonded particles. Experimental investigations have shown that bonded particle use leads to a better surface finish, whereas the material removal rate increases with unbonded MAPs. Figures 10.1a and 10.1b depict the schematic of unbonded and bonded abrasives to show the difference between them. The finishing efficacy of both bonded and unbonded abrasives has been shown in Figure 10.1c.

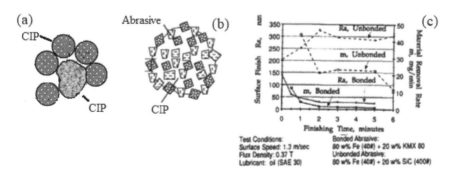

FIGURE 10.1 (a) Schematic of unbonded carbonyl iron particle (CIP) and abrasive (b) bonded CIP and abrasive and (c) finishing efficacy of bonded and unbonded particles.

Source: [5].

Advancement in MFAF Processes

The abrasive media used in all the variants are of two types: (i) dry mixture of abrasive with CIP, and (ii) mixture of abrasive, CIP, and carrier fluid. The mixture of CIP and abrasive in carried fluid makes the rheological property of the suspension controllable through an external magnetic field. In general, the mixture of CIPs and oil-based suspension is termed magnetorheological (MR) fluid. So, all the finishing processes which use the CIPs, abrasives, and oil-based suspension as a finishing media are known as MR based finishing processes, namely, MRF, MRAFF, etc. However, if the dry mixture of abrasive and CIP is used, the process is named as MFAF process. In a strong magnetic field, the fluid operates as solid; however, it acts as a fluid after removing the magnetic field. It may be observed from Figure 10.1b that the abrasive particle present at the end of the flexible brush performs the required finishing. Due to this, after a specific time of finishing, these abrasive particles' edges become blunt and require replenishment to maintain their efficacy. Therefore, the abrasive media needs to be changed after some time of finishing maintaining the efficacy of the process. So, due to the fluidic behaviour of MR fluid (wit out magnetic field), replenishment of the abrasive media becomes easy. Also, to increase efficacy, further etchants are used as the carrier fluid, which causes softening of surfaces easily removal.

The utility of most of the conventional finishing processes has been limited to flat surfaces. However, due to the use of the flexible brush, MFAF has been very effective in finishing complex surfaces. Due to this, various forms of MFAF processes have been developed to finish different surfaces. This chapter discusses different MFAF processes like BEMRP, MFP (used for complex shapes and surfaces), etc. Moreover, literature on experimental investigations carried out for each variant has also been discussed to understand the physics of each process.

10.2 BASICS OF MFAF PROCESS

Apart from so much variation in the setup used in all MFAF processes, the primary mechanism stays the same. Also, in all the variants, common process parameters govern the MFAF processes' efficacy. To better explain the basics, the process has been divided into process mechanisms and process parameters.

10.2.1 MECHANISM OF MFAF PROCESSES

As stated in the introduction, all of these MFAF processes use a flexible brush with embedded abrasives to remove the roughness present on the work surface. This flexible brush uses a mixture of abrasive particles and ferromagnetic material (mostly CIP) powder with or without the fluid medium. When the magnetic field is employed in the mixture, a chainlike structure is formed (refer to Figure 10.2a), and the strength and stiffness of this chain like structure are dependent on the applied magnetic field strength [6, 10]. The abrasive particle present in the mixture gets entangled between the chains and forms a flexible magnetic abrasive brush (refer to Figure 10.2a). Further, a relative motion (linear or circular) is provided to the flexible brush for executing the rubbing operation, which further results in a finishing operation (refer to Figure 10.2b).

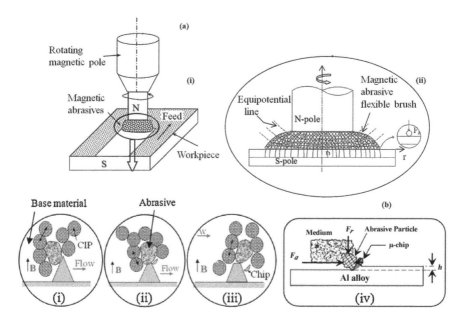

FIGURE 10.2 (a) Schematic of flexible brush in MFAF process, (b) material removal mechanism in MFAF processes.

Source: [6, 11].

The stiffness and strength of this brush may be modified by increasing or decreasing the applied magnetic field strength, which can affect the finishing process efficacy. The different stiffness of the flexible brush is very effective in finishing surfaces with different hardness.

10.2.2 Process Parameters in the MFAF Process

As in the schematic, the process appears to be simple, and there are only a few process parameters that affect the process efficacy like magnetic field strength, permeability of the mixture (of abrasive and CIP), size of particle in the mixture, and the relative velocity of abrasive brush and the work surface. In this section the effect of each process parameter is discussed in brief.

  a. *Effect of rubbing speed of particles on finishing efficacy:*

  It is observed from the primary research outcomes that as rubbing speed increases, the finishing efficacy first increases up to a specific value and then starts decreasing after a particular value of rubbing speed (refer to Figure 10.3a). It is because of the weak holding force of the flexible brush with which the abrasive particles are held between them. With the increase in rubbing velocity, the particles come out of the flexible brush and start separating from the tool surface, which further causes a reduction in finishing

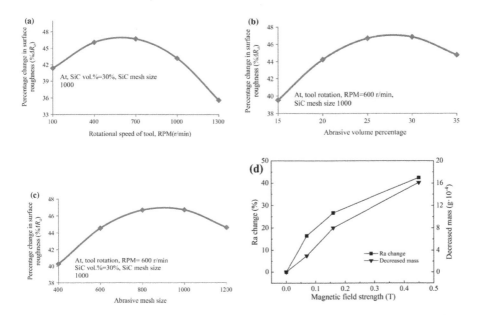

FIGURE 10.3 Effect of (a) tool rotational speed, (b) abrasive concentration, (c) abrasive size, (d), magnetic field strength.

Source: [14, 16].

efficacy [7, 12–14]. However, the holding strength of flexible brush is possibly be increased by enhancing the magnetic field strength, which further results in higher finishing efficacy at higher rubbing velocity.

b. *Effect of abrasive media on finishing efficacy:*

It has been found in the prior study that the abrasive concentration (in a mixture of CIP and abrasive) also has a significant effect on the finishing efficacy. It has been found that the finishing efficacy first increases to a particular value then decreases (refer to Figure 10.3b) after increasing the abrasive concentration. The researchers concluded that as abrasive concentration increases, the amount of cutting sites available for removing the asperities enhances and causes an improved finishing efficacy. However, too much abrasive concentration results in lower relative permeability of mixture, resulting in a lesser holding force of flexible chain and finishing efficacy. Apart from the abrasive concentration, its size also plays a crucial role in the finishing process. It has been found in the literature that with the rise in particle size, the process efficacy first increases and then decreases (refer to Figure 10.3c). However, the small abrasive particle size is recommended to surface with lower surface roughness.

In magnetic abrasive finishing (MAF), the outputs (i.e., material removal and surface finish) depend mainly on the process factors like work rotational speed and working gap [8]. The working gap in MAF is the distance of the workpiece from the magnets to construct the flexible magnetic abrasive brush. Besides, Chang et al. [9] also considered the finishing time effect on material removal and surface finish.

c. *Effect of magnetic field on finishing efficacy:*

The prior section has already discussed that the application of an external magnetic field can control the efficacy of MFAF processes [15]. Experimental investigations have shown that the finishing efficacy of the process improves with the rise in the magnetic field (refer to Figure 10.3d). The researchers have generated a physics-based model to relate the magnetic field strength to the polishing pressure, which governs the finishing efficacy. It can be seen from the relation that the polishing pressure is proportionate to the square of the magnetic field strength [16]. The magnetic field strength can also be altered by modifying the tool and work surface gap.

10.3 HYBRID MAGNETIC FIELD ASSISTED FINISHING PROCESSES

10.3.1 CHEMO-MECHANICAL MAGNETORHEOLOGICAL FINISHING

In order to enhance the efficacy of the MRF, it has been integrated with chemical polishing (CMP) to develop the chemo-mechanical magnetorheological finishing (CMMRF) process [17]. The CMMRF process possesses the vital features of both the MRF and CMP process and can be used to finish various materials like silicon, copper, etc. Figures 10.4a and 10.4b depict the schematic and finishing mechanism initial of the CMMRF process. Through CMP, the top layer of the process was softened, which is further removed with lesser finishing forces. Due to the oxidation of top surface hard and brittle materials like SiO_2, Al_2O_3 can also be finished with the CMMRF process. Experimental results have shown that CMMRF can also be used to finish soft and ductile material up to the nanometre level of surface roughness.

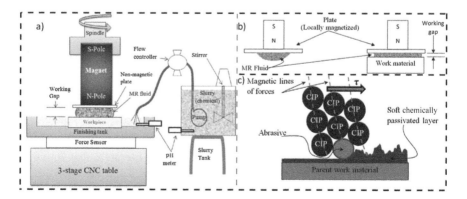

FIGURE 10.4 (a) Schematic of CMMRF process and (b) Material removal mechanism of CMMRF process.

Source: [18].

10.3.2 Ultrasonic Assisted Magnetic Abrasive Finishing

It is another form of hybrid MAF process. In this process, ultrasonic vibration motion is over laid with the conventional MAF process. Like the traditional MAF process, the ultrasonic assisted magnetic abrasive finishing (UAMAF) process also utilizes abrasive and CIP mixture for the finishing process. The aid of ultrasonic vibration increases the relative velocity between the work surface and abrasives, which causes higher interaction between abrasive and surface asperities resulting in higher finishing efficacy [19–21]. Figure 10.5 shows a schematic representation of UAMAF; the setup includes a power source, a piezoelectric transducer with a horn attached, and a power generator that produces electrical signals at 20 kHz frequency. The normal force generated on CIP due to the magnetic field provides the required indention to the abrasive. Experimental results have shown that the assistance reduces the time of finishing needed for hours to minutes.

Few literatures also reported the improved performance of the process by taking combined effect of ultrasonic vibration and chemical polishing assistance with the MAF and MRF process. Experimental results have also shown that this combined effect is very influential on the finishing efficacy of the process irrespective of workpiece material property.

10.3.3 Electrolyte Magnetic Abrasive Finishing

Electrolyte magnetic abrasive finishing (EMAF) is another variant of the MAF process, wherein the electrolytic process is utilized to enhance the quality of the surface finish. In the EMAF process, an oxide layer is produced for the material removal from the surface easily [23].

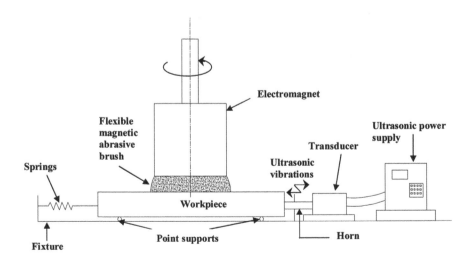

FIGURE 10.5 Schematic of UAMAF process.

Source: [22].

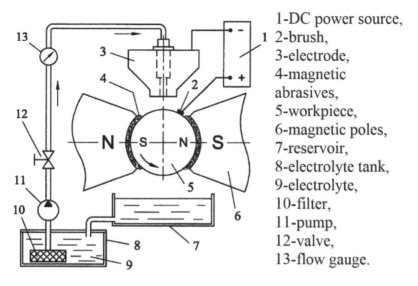

1-DC power source,
2-brush,
3-electrode,
4-magnetic abrasives,
5-workpiece,
6-magnetic poles,
7-reservoir,
8-electrolyte tank,
9-electrolyte,
10-filter,
11-pump,
12-valve,
13-flow gauge.

FIGURE 10.6 Schematic representation of EMAF.
Source: [23].

The schematic view of the EMAF process is depicted in Figure 10.6. The EMAF is a hybrid process in which mechanical and electrochemical energies are utilized for material removal. As illustrated in Figure 10.6, a provision of electrolytes is provided to the conventional MAF experimental setup.

During EMAF, the electrolyte is filled in the workpiece and electrode gap. Subsequently, the workpiece and electrode are linked to the anode and cathode of the DC power sources. In this process, the magnetic field of the working gap affects the formation of a flexible magnetic abrasive brush for the application of abrasion pressure caused by the magnetic forces. In addition to this, an electrochemical reaction creates a passive layer over the work surface during EMAF. Consequently, the softened passive layer provides an easier material removal by the MAF process.

During MAF, the magnetic field spreads in the electrode gap; however, the electric field assists the electrolytic process. In addition, the magnetic and electric fields are perpendicular to each other. Figure 10.7a shows the electrolytic process without the magnetic field, wherein electric force directs the negatively charged ions to drive towards the anodic surface of the metal in a linear direction to get further reacted. It results in a highly concentrated electric force at the crest of the surface, thereby forming a thicker passive layer at the crests rather than the valleys of the surface. While, if the magnetic field is introduced in the electrolytic process as depicted in Figure 10.7b, Lorentz force affects the path of negatively charged ions to move in a cycloidal manner towards the anodic surface further to enhance the collision of electrolytic ions and unionized electrolyte, thereby improving the electrochemical reaction of an electrolytic process. This process helps in creating a thicker passive

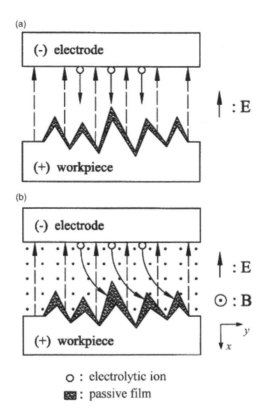

FIGURE 10.7 The effect of (a) no magnetic field and (b) magnetic field during the electrolytic process.

Source: [23]

layer to the unattended valleys, which were left by the electrolytic process under the influence of no magnetic field. Such a thicker passive layer of the work-surface yields a better super finishing of the surface.

10.4 DIFFERENT MFAF PROCESSES FOR NON-PLANAR SURFACES

10.4.1 Magnetorheological Finishing Process

High precision glass based optical lenses require a very high surface finish, and while finishing, crack gets generated during the finishing operation. Also, these lenses are of either convex or concave shape, which is hard to finish with the use of traditional processes. In order to overcome this bottleneck, researchers have developed a finishing technology called magnetorheological finishing (MRF) operation [24]. The setup consists of an electromagnet, a ferromagnetic wheel, workpiece holding setup, pump, nozzle, and MR fluid conditioning unit. The workpiece holding setup has the

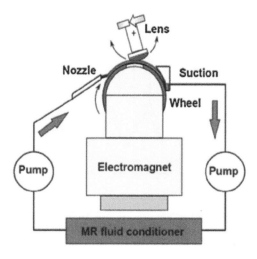

FIGURE 10.8 Schematic of MRF process.

Source: [6].

dexterity to orient the workpiece to perform the finishing operation at any desired portion of the lens. The electromagnet is used to stiffen the MR fluid for finishing. The ferromagnetic wheel is used for replenishing the MR fluid. Other parts are for supplying and cleaning the MR fluid.

In this process, a suspension of abrasive in magnetorheological (MR) fluid is employed as a finishing media. This MR fluid is a suspension of CIPs spread in a non-magnetic fluid (i.e. water, silicone oil, etc.). MR fluid behaves similar to the Newtonian fluid in the lack of magnetic environment; however, in the presence of a magnetic field, it acts as a non-Newtonian fluid. The schematic diagram of MRF based lens finishing process is presented in Figure 10.8. The distribution of CIPs in MR fluid in the presence and absence of magnetic field is depicted in Figures 10.9a and 10.9b. Figure 10.9c represents that the interaction between finishing media and the workpiece (Lens). The magnetic field strength governs the stiffness of CIP chains which directly controls the penetration force. Because of the penetration force, the abrasives indent into the work surface, and by moving the abrasive media for material removal in micro-/nanosized chips, thereby resulting in nanofinishing. The upper limit of the magnetic field up to which the efficacy keeps improving depends on magnetic saturation of the MR fluid. The MRF finishing process can produce a surface finish up to 10 nm peak to valley height on optical lenses [24].

10.4.2 Magnetorheological Abrasive Flow Finishing

Magnetorheological abrasive flow finishing (MRAFF) was developed to fine finish the inner cylindrical surfaces with complex geometries [10]. This process can be seen as an amalgamation of abrasive flow finishing and magnetorheological finishing

Advancement in MFAF Processes

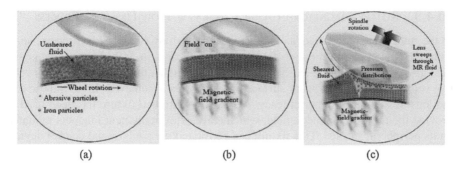

FIGURE 10.9 CIP distribution in MR fluid and abrasive suspension in the (a) absence, (b) presence,(c) during interaction with the workpiece in the presence, of the magnetic field.

Source: [6].

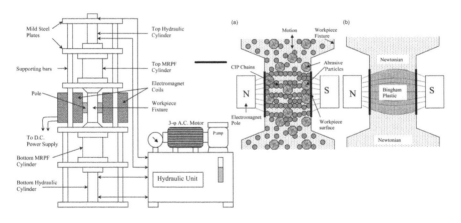

FIGURE 10.10 (a) Schematic representation and (b) mechanism of material removal of MRAFF process.

Source: [10].

process. In this process, MR fluid oscillates through the workpiece so that the abrasive media can carry out the finishing on the inner surface.

Additionally, the application of an external magnetic field at the peripheral surface of the workpiece results in the formation of chain structure in MR fluid with abrasive entangled in between. Because of the applied magnetic field, these chains exercise the normal force on the inner work surface, increasing finishing pressure to further improves process efficacy.

Due to the existence of a controllable magnetic field, the performance of the abrasive flow finishing process can be altered to obtain the desired surface finish. However, owing to the application of an external magnetic field, this process is also limited to pipes and shells. The schematic of the MRAFF process is represented in Figure 10.10.

10.4.3 ROTARY MAGNETORHEOLOGICAL ABRASIVE FLOW FINISHING PROCESS

In the rotary magnetorheological abrasive flow finishing (RMRAFF) process, apart from the reciprocating motion of MR fluid (as present in MRAFF), the magnetic field also rotates about the axis of the cylinder through a fixture [25]. A higher relative velocity between the work surface and abrasives is achieved through these two motions, resulting in higher efficacy. A specially designed fixture has been used to provide the rotating magnetic field, which consists of permanent magnets mounted on its peripheral surface. This fixture can be rotated via a motor and control device. Due to this rotating magnetic field, the flexible brush formed inside the workpiece revolves and finishes the inner work surface. Consequently, the inner work surface gets finished, and high surface quality is obtained. However, similar to the MRAFF process, this process is also limited to pipes and shells because of the application of an external magnetic field. Figures 10.11(a) and 10.11(b) show the schematic of RMRAFF and the finishing mechanism involved.

10.4.4 MAGNETIC FLOAT POLISHING

Magnetic float polishing (MFP) is another variant of MFAF processes used to finish the spherical surfaces. Figure 10.12 presents the schematic diagram of MFP process. Despite the complex setup used in the process, the material removal mechanism remains almost the same as all MFAF processes. The lowest average surface roughness obtained with the MFP process on the balls made up of hard ceramic material is 4 nm. Apart from improvement in surface finish, the sphere form also improved and further resulted in 150–200 nm of sphericity. The experimental investigations revealed that MFP could be used to finish ceramic balls and bearing rollers without scratch marks and pits [26–27]. This mechanism is established on the ferro-hydrodynamic behaviour of magnetic fluid that causes levitation in the non-magnetic float and abrasives. This force of levitation is in proportion to the controlled applied magnetic field gradient. A stack of strong electromagnets is mounted in alternate

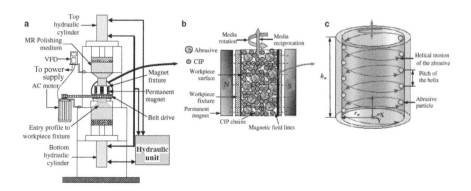

FIGURE 10.11 (a) Schematic of RMRAFF process and (b) material removal mechanism of RMRAFF process.

Source: [25].

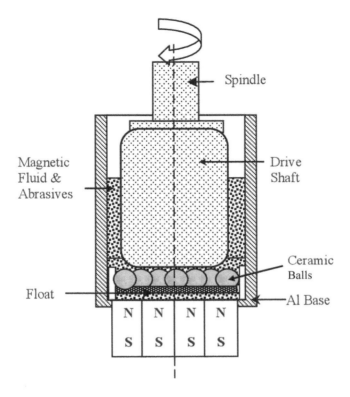

FIGURE 10.12 Schematic of MFP process.

Source: [6].

order in north and south poles orientation underneath the fluid chamber. The higher magnetic field area attracts the ferro-fluid, and a buoyant force is applied to non-ferro materials to force them towards the reduced magnetic field [6]. All the non-magnetic materials experience the buoyant force, and this force results in the levitation of abrasive grains, ceramic balls, and acrylic float within the chamber. In order to touch and pressurise the balls, the driveshaft is pushed downwards. Then the surface finishing of spheres is achieved via abrasives and balls action. Tani and Kawata successfully demonstrated that the MFP efficacy on Si_3N_4 balls used high-speed hybrid bearing for jet turbines of aircraft [28].

10.4.5 Ball End Magnetorheological Polishing Process

Super finishing surfaces with complex features (like die for coin manufacturing) are considered difficult to achieve, especially for tough materials. In order to satisfy this issue, the ball end magnetorheological polishing (BEMRP) process was developed [29]. In this process, an abrasive laden ball end shaped stiffened MR polishing (MRP) tip is formed at the finishing tip of the rotating fixture. Apart from finishing complex 3D morphologies, the BEMRP process is also very effective in finishing flat surfaces. During finishing, the flow of MR fluid at the tool tip surface is intermittent. In the BEMRP process, the ball end of the tool is utilized to perform the finishing operation. Like other MFAF processes, the finishing media must be replenished to maintain efficacy. The conditioning of finishing media in the BEMRP process is done intermediately by replenishing the flexible MR ball. Because of the round shape of the finishing tip, the BEMRP process can easily be 3D surface profiles with a uniform surface finish. Due to this flexibility, the BEMRP process has found its applicability on constricted surfaces such as in-depth pockets, grooves, and inclined surfaces of a workpiece. Finishing this type of surface is difficult through other MFAF based processes like MRF, MRAFF, etc. Figure 10.13 depicts the schematic diagram of the BEMRP experimental setup.

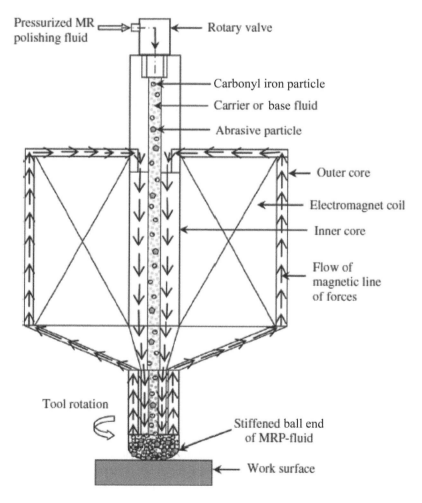

FIGURE 10.13 Schematic of BEMRP process.
Source: [29].

10.4.6 Magnetorheological Jet Finishing Process

The magnetorheological jet finishing (MRJF) process was created to finish freeform dome optics, cavities, and steep concave optics, among other things [30]. In the MRJF process, a jet of MR polishing fluid is thrown into the internal surface or geometry of the workpiece. After coming out of the nozzle, the MR polishing fluid jet is subjected to an axial magnetic field [30]. Due to this, CIP particles get accelerated towards the work surface, and the entangled abrasives can perform the finishing operation in the deep internal cavities. The surface roughness value is improved from 0.54 µm to 0.04 µm [30]. A schematic view of magnetic abrasive jet finishing (MRJF) is demonstrated in Figure 10.14a. The material removal in MRJF is governed by impacting the abrasive particle's kinetic energy. As shown in Figure 10.14b, the jet of MR fluid loses its coherence coming out of the nozzle without the magnetic field. In contrast, a stable jet of MR fluid that comes out of the nozzle can be observed with the presence of a magnetic field (refer to Figure 10.14c). This stable jet of MR fluid is used to finish the deep cavity surface of workpiece materials.

10.4.7 Magnetorheological Abrasive Honing Process

The magnetorheological abrasive honing (MRAH) process is employed for finishing the internal surfaces of through and blind holes [31]. The tool motion of the MRAH process is the same as the conventional honing process. In the MRAH process, a specially designed fixture with an electromagnet is used as a finishing tool (refer to Figure 10.15a). For executing the MRAH process, the fixture/tool with abrasive media (CIP brush with embedded abrasives) on its periphery is allowed a to and fro motion with the help of a plunger, as represented in Figure 10.15b. Experimental results have shown that the MRAH process is only efficacious on non-ferrous materials. The

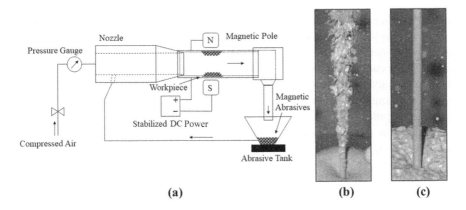

FIGURE 10.14 (a) Schematic of MRJF process, and the jet of MR fluid (b) without and (c) with the magnetic field.

Source: [30].

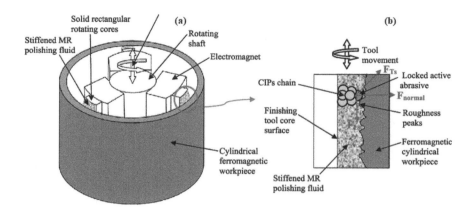

FIGURE 10.15 (a) Schematic of MRAH process and (b) material removal mechanism in MRAH process.

Source: [32].

experimental results have also shown that the finishing efficacy reforms with the magnitude of current supplied to the electromagnet.

Due to the large size of electromagnets, the aforesaid tool cannot be used for smaller size holes. However, a few researchers have also developed a permanent magnet based honing tool for smaller sized holes to overcome this bottleneck. Despite using a permanent magnet, the developed tool is proficient in producing a variable magnetic field. The finishing efficacy of the developed tool was also verified on the SS 304 pipes.

10.4.8 Magnetic Abrasive Finishing of Cylindrical Parts

In the modern era, new materials have been evolved to serve a variety of applications. These materials are difficult to cut; so, there exists a demand for the alternate manufacturing process to machine/finish these materials without amending the quality of finish and accuracy. Conventional methods like grinding are restricted to produce such quality accuracy and finish with minimal surface defects. Another way of producing such quality accuracy and surface finish is the polishing methods. However, the polishing may lead to a rise in the cost of the part.

MAF is an abrasive based finishing process. It is employed with a promise to produce mirror like, the highly accurate surface finish of the parts without any damage, microcracks to the finished surface. This technique has been evolved as one of the crucial finishing techniques for hard materials such as advanced ceramics (i.e., silicon nitrides) [5]. MAF is a magnetic field assisted process that is proficient in producing (i) high finish and accuracy of the surface, (ii) no surface damages during the finishing operations owing to very low cutting forces, and (iii) very high operating spindle speeds in comparison to the conventional techniques. Other than this, various characteristics are associated with MAF processes, such as self-sharpening,

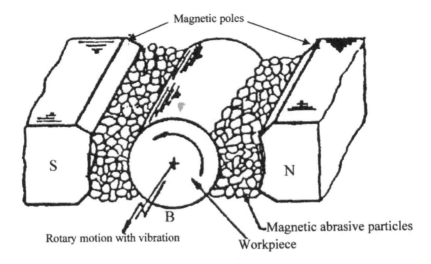

FIGURE 10.16 Schematic representation of CMAF.
Source: [8].

controllability, self-adaptability, and no requirement of dressing/truing to the machining tool [9]. When utilized for the cylindrical components/parts, MAF is known as cylindrical magnetic abrasive finishing (CMAF).

The major elements of the CMAF process are the magnetic abrasive particles (MAPs), magnetic poles (either permanent or electromagnet), and provisions of workpiece rotations and vibrations [33]. Figure 10.16 presents the schematic view of the CMAF process. Apart from this, the CMAF process is also greatly influenced by the size of the ferromagnetic/abrasive particles, finishing time, type of MAPs (bonded, sintered, and unbonded) [5, 8–9, 33]. The CMAF finds its applicability in several areas due to its versatility and capability to produce mirror like surface finish to difficult-to-finish material with minimum wastage. It is also employed to finish the inner and outer work surfaces. The forthcoming section discusses the variety of applications of the CMAF process.

10.4.9 INTERNAL MAGNETIC ABRASIVE FINISHING

Internal magnetic abrasive finishing (IMAF) can be utilized to finish internal surfaces like tubes, gas, and liquid piping systems [34–36]. Figures 10.17a and 10.17b show the schematic representation with stationary poles and the abrasive behaviour of the IMAF system. The abrasive behaviour influences the finishing characteristics of the inner surface of tube/pipe concerning the inner surface of the tube/pipe. Figure 10.17b depicts the four stages of this abrasive behaviour of MAF.

All the abrasives are kept magnetically in the finishing zone and show the clean rotation by following the magnetic field movement along the inner surface to improve further the workpiece's internal surface finish (refer to Figure 10.17a). Then,

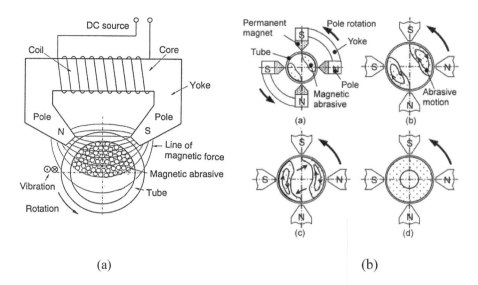

FIGURE 10.17 Schematic representation of internal MAF (a) using stationary pole system, and the (b) abrasive behaviour during the process.

Source: [34, 35].

abrasives undergo different stages, wherein the sequentially weaker magnetic force is obtained; this abrasive behaviour may be inferred from Figures 10.17b (stages b-d).

Figure 10.17b presents a stage of abrasives that rotate against the inner surface and self-displace under the abrasive mass. Self-displacement of the abrasive explains the motion of the cutting edges of the abrasives under the abrasive mass to cause an increased material removal further. In stage 3, the magnetic field merely holds the fraction of abrasive mass in the finishing zone, thereby dispersing the abrasives away from the finishing zone erratically to strike and cut the work surface. Therefore, a difference in the modified surface is evident from the stages of Figures 10.17a and 10.17b, and a fiery abrasive contact against the work surface results in noteworthy enhanced material removal. Further, the magnetic field could not contain the abrasives in the finishing zone due to a reduction in the magnetic force. In this stage, the abrasive sticks to the rotating work surface due to the friction irrespective of the movement of the magnetic field to further eliminate the relative motion between the abrasives and the inner work surface and to retain the initial work surface of the material. Furthermore, it can be inferred that the excessive supply of the abrasive and high rotational magnetic field speed may result in an unstable condition of abrasive in the finishing zone.

10.5 CONCLUSIONS

Based on the literature review performed on MFAF processes, the following conclusions can be drawn.

- BEMRP process is suitable mainly for finishing complex features on a surface like engraving and protrusion on a die.

- MRF process is the most efficacious process for finishing concave and convex lenses.
- RMRAFF process is highly suitable to finish the inner work surface of the bearing race.
- CMAF process is mainly suitable to finish the outer work surface of the cylinder like pistons.
- All MFAF processes have very low MRR, so it is not recommended for industries to produce low-cost products in high quantities.

10.6 LIMITATIONS AND FUTURE SCOPE OF MFAF PROCESS

10.6.1 LIMITATIONS

Based on the literature review performed on MFAF processes it can be concluded that the finishing rate of all its variants are quite low and can not be used for bulk production industries. Additionally, finishing of products made of ferromagnetic material using MAF based processes is very difficult.

10.6.2 FUTURE SCOPE

From the literature review it can be concluded that the despite its slow rate of MFAF processes, the surface quality achieved has all the properties required for metallic mirrors which is critical for telescopes. Apart from high surface quality, the variants of MFAF processes cover almost all types of surfaces, so products of any shape can be finished. Additionally, the setup and processing costs of MFAF-based processes are low compared to other super finishing operations which makes it an economical option.

REFERENCES

[1] J. Long *et al.*, "Fused silica contamination layer removal using magnetic field-assisted finishing," *J. Am. Ceram. Soc.*, vol. 103, no. 5, pp. 3008–3019, 2020.

[2] P. Y. Wu, M. Hirtler, M. Bambach, and H. Yamaguchi, "Effects of build- and scan-directions on magnetic field-assisted finishing of 316L stainless steel disks produced with selective laser melting," *CIRP J. Manuf. Sci. Technol.*, vol. 31, no. 2019, pp. 583–594, 2020.

[3] C. Wang, C. F. Cheung, L. T. Ho, K. L. Yung, and L. Kong, "A novel magnetic field-assisted mass polishing of freeform surfaces," *J. Mater. Process. Technol.*, vol. 279, p. 116552, 2020.

[4] M. Jiang, P. Examiner, D. A. Scherbel, A. Examiner, and P. J. Hoffmann, "United States patent (19)," vol. 2, no. 19, 1999.

[5] M. Fox, K. Agrawal, T. Shinmura, and R. Komanduri, "Magnetic abrasive finishing of rollers," *CIRP Ann.—Manuf. Technol.*, vol. 43, no. 1, pp. 181–184, 1994.

[6] V. K. Jain, "Magnetic field assisted abrasive based micro-/nano-finishing," *J. Mater. Process. Technol.*, vol. 209, no. 20, pp. 6022–6038, 2009.

[7] P. Kala and P. M. Pandey, "Experimental study on finishing forces in double disk magnetic abrasive finishing process while finishing paramagnetic workpiece," *Procedia Mater. Sci.*, vol. 5, pp. 1677–1684, 2014.

[8] V. K. Jain, P. Kumar, P. K. Behera, and S. C. Jayswal, "Effect of working gap and circumferential speed on the performance of magnetic abrasive finishing process," *Wear*, vol. 250–251, no. PART 1, pp. 384–390, 2001.

[9] G. W. Chang, B. H. Yan, and R. T. Hsu, "Study on cylindrical magnetic abrasive finishing using unbonded magnetic abrasives," *Int. J. Mach. Tools Manuf.*, vol. 42, no. 5, pp. 575–583, 2002.

[10] S. Jha and V. K. Jain, "Design and development of the magnetorheological abrasive flow finishing (MRAFF) process," *Int. J. Mach. Tools Manuf.*, vol. 44, no. 10, pp. 1019–1029, 2004.

[11] V. K. Jain, "Advanced (non-traditional) machining processes," *Mach. Fundam. Recent Adv.*, pp. 299–327, 2008.

[12] R. S. Mulik and P. M. Pandey, "Experimental investigations and optimization of ultrasonic assisted magnetic abrasive finishing process," *Proc. Inst. Mech. Eng. Part B J. Eng. Manuf.*, vol. 225, no. 8, pp. 1347–1362, 2011.

[13] G. C. Verma, P. Kala, and P. M. Pandey, "Experimental investigations into internal magnetic abrasive finishing of pipes," *Int. J. Adv. Manuf. Technol.*, vol. 88, no. 5–8, pp. 1657–1668, 2017.

[14] S. A. Sirwal, A. K. Singh, and S. K. Paswan, "Experimental analysis of magnetorheological finishing of blind hole surfaces using permanent magnet designed tools," *J. Brazilian Soc. Mech. Sci. Eng.*, vol. 42, no. 3, 2020.

[15] Z. Fan, Y. Tian, Q. Zhou, and C. Shi, "A magnetic shear thickening media in magnetic field—assisted surface finishing," *Proc. Inst. Mech. Eng. Part B J. Eng. Manuf.*, vol. 234, no. 6–7, pp. 1069–1072, 2020.

[16] W. Song, Z. Peng, P. Li, P. Shi, and S. B. Choi, "Annular surface micromachining of titanium tubes using a magnetorheological polishing technique," *Micromachines*, vol. 11, no. 3, 2020.

[17] V. K. Jain, P. Ranjan, V. K. Suri, and R. Komanduri, "Chemo-mechanical magnetorheological finishing (CMMRF) of silicon for microelectronics applications," *CIRP Ann.—Manuf. Technol.*, vol. 59, no. 1, pp. 323–328, 2010.

[18] P. Ranjan, R. Balasubramaniam, and V. K. Jain, "Analysis of magnetorheological fluid behavior in chemo-mechanical magnetorheological finishing (CMMRF) process," *Precis. Eng.*, vol. 49, pp. 122–135, 2017.

[19] A. Misra, P. M. Pandey, U. S. Dixit, A. Roy, and V. V. Silberschmidt, "Modeling of finishing force and torque in ultrasonic-assisted magnetic abrasive finishing process," *Proc. Inst. Mech. Eng. Part B J. Eng. Manuf.*, vol. 233, no. 2, pp. 411–425, 2019.

[20] A. Misra, P. M. Pandey, and U. S. Dixit, "Modeling and simulation of surface roughness in ultrasonic assisted magnetic abrasive finishing process," *Int. J. Mech. Sci.*, vol. 133, pp. 344–356, 2017.

[21] A. Misra, P. M. Pandey, and U. S. Dixit, "Modeling of material removal in ultrasonic assisted magnetic abrasive finishing process," *Int. J. Mech. Sci.*, vol. 131–132, pp. 853–867, 2017.

[22] R. S. Mulik and P. M. Pandey, "Mechanism of surface finishing in ultrasonic-assisted magnetic abrasive finishing process," *Mater. Manuf. Process.*, vol. 25, no. 12, pp. 1418–1427, 2010.

[23] B. H. Yan, G. W. Chang, T. J. Cheng, and R. T. Hsu, "Electrolytic magnetic abrasive finishing," *Int. J. Mach. Tools Manuf.*, vol. 43, no. 13, pp. 1355–1366, 2003.

[24] W. Kordonski and S. Jacobs, "Model of magnetorheological finishing," *J. Intell. Mater. Syst. Struct.*, vol. 7, no. 2, pp. 131–137, 1996.

[25] M. Das, V. K. Jain, and P. S. Ghoshdastidar, "Nanofinishing of flat workpieces using rotational-magnetorheological abrasive flow finishing (R-MRAFF) process," *Int. J. Adv. Manuf. Technol.*, vol. 62, no. 1–4, pp. 405–420, 2012.

[26] N. Umehara, T. Kirtane, R. Gerlick, V. K. Jain, and R. Komanduri, "A new apparatus for finishing large size/large batch silicon nitride (Si3N4) balls for hybrid bearing applications by magnetic float polishing (MFP)," *Int. J. Mach. Tools Manuf.*, vol. 46, no. 2, pp. 151–169, 2006.

[27] M. Raghunandan, N. Umehara, A. Noori-Khajavi, and R. Komanduri, "Magnetic float polishing of ceramics," *J. Manuf. Sci. Eng. Trans. ASME*, vol. 119, no. 4, pp. 520–528, 1997.

[28] Y. Tani and K. Kawata, "High-efficient fine finishing process using magnetic fluid," *CIRP Ann.*, vol. 33, pp. 7–10, 1984.

[29] A. Kumar Singh, S. Jha, and P. M. Pandey, "Design and development of nanofinishing process for 3D surfaces using ball end MR finishing tool," *Int. J. Mach. Tools Manuf.*, vol. 51, no. 2, pp. 142–151, 2011.

[30] W. I. Kordonski, A. B. Shorey, and M. Tricard, "Magnetorheological jet (MR Jet™) finishing technology," *J. Fluids Eng. Trans. ASME*, vol. 128, no. 1, pp. 20–26, 2006.

[31] A. Sadiq and M. S. Shunmugam, "Investigation into magnetorheological abrasive honing (MRAH)," *Int. J. Mach. Tools Manuf.*, vol. 49, no. 7–8, pp. 554–560, 2009.

[32] S. K. Paswan, T. S. Bedi, and A. K. Singh, "Modeling and simulation of surface roughness in magnetorheological fluid based honing process," *Wear*, vol. 376–377, pp. 1207–1221, 2017.

[33] K. B. Judal, V. Yadava, and D. Pathak, "Experimental investigation of vibration assisted cylindrical-magnetic abrasive finishing of aluminum workpiece," *Mater. Manuf. Process.*, vol. 28, no. 11, pp. 1196–1202, 2013.

[34] H. Yamaguchi and T. Shinmura, "Study of the surface modification resulting from an internal magnetic abrasive finishing process," *Wear*, vol. 225–229, no. 1, pp. 246–255, 1999.

[35] H. Yamaguchi and T. Shinmura, "Study of an internal magnetic abrasive finishing using a pole rotation system. Discussion of the characteristic abrasive behavior," *Precis. Eng.*, vol. 24, no. 3, pp. 237–244, 2000.

[36] H. Yamaguchi and T. Shinmura, "Internal finishing process for alumina ceramic components by a magnetic field assisted finishing process," *Precis. Eng.*, vol. 28, no. 2, pp. 135–142, 2004.

Index

0–9

3D printer emissions and air quality control for health, 105

3D printing applications, 103

3D printing responds to pandemic, 102

A

additive manufacturing materials, 83

additive manufacturing processes, 80

advances in environmental protection, 105

advantages of AHP method, 110

advantages of thermal plasma processing of materials, 152

aerospace, 90

application of rotary ultrasonic machining in the field of biomedical, 160

applying AHP method, 109

autoclave cure prepreg process, 124

B

ball end magneto rheological polishing process, 185

basics of MFAF process, 175

biomaterials, 91

buckling analysis, 64

buckling criterion, 60

C

CAD geometry with hexagonal perforations, 60

carbon nanotubes, 122

case study, 16

ceramics, 86

characterization of nanofluids, 19

chemo-mechanical magneto rheological finishing, 178

conventional cooling, 9

conventional cutting fluids, 4

cryogenic cooling, 12

cutting fluids, 4

D

data analysis, 33, 111

design of experiments, 133

details of electrodes material, 31

different MFAF processes for non-planar surfaces, 181

direct metal laser sintering technology, 58

dry machining, 9

E

effect of abrasive media on finishing efficacy, 177

effect on drill design on output response, 168

effect on feed rate on output response, 165

effect on human health, 105

effect on input parameters on output response, 164

effect of magnetic field on finishing efficacy, 178

effect of process parameters on EWR, 44

effect on rotational speed on output response, 164

effect of rubbing speed of particles on finishing efficacy, 176

electrolyte magnetic abrasive finishing, 179

electro-mechanical engineering, 179

F

framework of criteria and sub-criteria, 106

fundamentals of plasma, 149

fused deposition modelling (FDM), 81

G

genetic algorithm-optimization, 47

green 3D printing, 103

H

hexagonal perforations, 53

hybrid magnetic field assisted finishing processes, 178

I

influence of control factors on MRR, 42

internal magnetic abrasive finishing, 189

L

limitation of the study and future scope, 169, 191

M

machining study, 21

magneto rheological abrasive flow finishing, 182

magneto rheological finishing process, 181

material testing methods, 133
mechanism of MFAF processes, 175
modal analysis and frequency response, 63
minimum quantity lubrication (MQL), 10
mundane or artificially produced plasmas, 151

N

nanofluids, 6
nanotechnology in composite aerospace
    structures, 124
nanotechnology in nanomaterial's
    manufacturing, 121
nanotechnology in product-based
    manufacturing, 122
non-conventional cooling, 9
numerical optimization, 142

O

occurrence of plasma in nature, 151

P

polymers, 88
process description, material and method, 131
process optimization, 47
process parameters in the MFAF process, 176

Q

quadratic equation, 140

R

regression analysis, 55

S

resin transfer molding process (RTM), 125
response surface of MRR, 37
response surface of EWR, 39
rotary magneto rheological abrasive flow
    finishing process, 184

selective laser sintering (SLS), 82
stability study, 17
steps for AHP method, 110
stereolithography (SLA), 82
structural analysis, 63
surface morphology, 46

T

techniques of cleaner machining, 8
three dimensional printing metals, 56
tool design, 31
tribological study, 20

U

ultrasonic assisted magnetic abrasive
    finishing, 179
ultrasonic atomization of cutting fluid, 15

V

vegetable oil-based cutting fluids, 5

W

wall angle, 136
wall thinning, 139